AIR DEFENCE AT SEA

Previous page:
A development that came too late to help the Falklands conflict: Sea King Mk 6 AEW
Crown Copyright

AIR DEFENCE AT SEA

Rear Admiral J. R. Hill

LONDON

IAN ALLAN LTD

First published 1988

ISBN 0 7110 1742 5

Published by Ian Allan Ltd, Shepperton, Surrey; and printed by Ian Allan Printing Ltd at their works at Coombelands in Runnymede, England

Acknowledgements

In some ways, *Air Defence at Sea* was always going to be an easier book to produce than *Anti-Submarine Warfare*, my first venture in this series. When at sea my trade had been in the above-water world, and it is less obscure than the underwater one anyway. Moreover, its instruments are more photogenic and pictures of them are generally more available.

None of this meant that I didn't need a lot of help in preparing the book. Air defence is a developing, sometimes swiftly developing, business, and my knowledge needed bringing up to date in many areas. Among those who helped to do this were Cdr Paul Canter, Lt Cdr Duggie Hamilton, Lt David Hazel, Cdr Mackenzie, Lt Cdr Rick Mitchell, Capt Peterson USN, Lt Cdr Michael Plumridge, Antony Preston, Capt John Tolhurst and Anna C. Urband.

There are two people to whom special thanks are due, partly for putting me in touch with some of those already mentioned. These are Adm Sir Julian Oswald and Capt Colin Cooke-Priest. They made sure I heard all I could properly hear, and with the right emphasis, from Flag staff and Maritime Tactical School respectively. They were able to do this because they were in the right place at the time, and I hope they will approve the book as reflecting their expertise.

Photographs are an important part of a book of this kind and I am indebted not only to those who have contributed to my photographic library in the past, but to a number who produced photographs specially for this issue: Russell Egnor of the US Department of the Navy, Mrs J. Hamer (VSEL), Cdr R. B. Rankin (BMARC), M. A. Tenot (Creusot Loire), Mlle Jocelyne Terrien (Aerospatiale), Philip M. Treloar (Ferranti), Michelle Utton (Thorn EMI) and Mr Willis of the Imperial War Museum. Several other organisations were most helpful in providing photographs and I hope all are properly acknowledged in the captions.

Finally, I should as always like to acknowledge the unfailing support of the Ian Allan Ltd team under Simon Forty.

Contents

Introduction

On Christmas Day 1914 a motley collection of British seaplanes, operating from an area in the German Bight, took off to attack the German Zeppelin sheds and units of the High Seas Fleet in Cuxhaven. Their parent units were three hastily adapted ships which had the job of launching them on to, and recovering them from, the water. They were independently navigated, their bomb loads were tiny, their communications varied between the minimal and the non-existent and the visibility was patchy.

Judged by results, the raid was not a success. Several aircraft did not arrive in the target area at all. Those that did liberally misidentified their targets — coasters and freighters being mistaken for cruisers — and nearly all the bombs fell wide of the mark. The main land target, a Zeppelin shed, was missed by a wide margin, although a fuel tank some distance away was blown up with spectacular results. Through navigational hazard and because of counter-attacks on their parent units by the Germans — ineffective in material terms but of deterrent effect — some aircraft did not return to their launching ships.

Yet this first attack from the air on warships was a remarkable achievement, occurring as it did only 11 years after the first powered flight was made by Orville Wright at Kitty Hawk. The advances made by aviation in the intervening years had indeed been astonishingly rapid; and that pace was to continue.

The speed of development was not confined to technical or material advance. In most industrialised countries a good deal of thought had been given to the use of aircraft in sea warfare; and sea warfare was, of course, a fashionable matter for discussion in those days when the Kaiser read the works of Mahan, 'Jacky' Fisher wielded such influence with the government of Britain, and Theodore Roosevelt sent the United States Navy round the world.

In Britain, characteristically, thinking about the use of the air weapon at sea — and about defence against it — was not officially sponsored but was left to individuals. There was no shortage of ideas, nevertheless; and there was considerable freedom of expression, typified by the formation of the Naval Society in 1912 and its *Naval Review* which first appeared in 1913. One of the first articles was a long piece starkly entitled 'Air Power' and it makes interesting reading today.

The author — anonymous, as were all contributors to *The Naval Review* at that time — saw roles for aircraft, even in their current state of development, as reconnaissance, anti-submarine work and attack on unarmoured ships. He added ' . . .as we develop the offensive capacity of the machines, so will they be able to attack larger and larger ships. . . to protect ships from an attack of this nature, we are now developing two classes of gun. . .', and went on to say that crews would need to be at action stations continuously, with consequent strain on personnel.

It is all stirring stuff. The author got quite a lot wrong, in his overconfidence in the material durability of floatplanes and his emphasis on airships for example, but the general accuracy of the forecasting and the assessment of its implications is remarkable.

Even more prophetic was the work of F. W. Lanchester, who in 1914 was confidently stating a staff requirement for a 'pontoon-ship' that almost exactly matched the first generation of aircraft carriers, and very accurately assessed the principles governing both the offensive and defensive in air warfare at sea. It is conceivable that had the ideas in these and other articles been fully taken up and pursued with the impetus that is achievable in wartime, even greater progress would have been made before 1918 and a much more potent Royal Naval Air Service evolved. For, truth to tell, very few air attacks were made on ships by either side in that war. Air effort at sea was concentrated on reconnaissance and observation with, on the British side, an emphasis on anti-submarine operations. The use of air power as an offensive measure against ships at sea was yet to come.

In the area of theory and experiment, at least, it

Above:
A German torpedo boat damaged in an air raid on Brugge harbour during World War 1.
Imperial War Museum (IWM Q58103)

Left:
The first operational aircraft carrier: aircraft ranged on deck in HMS *Furious* in preparation for an attack on the Zeppelin sheds at Tondern in 1918. The aircraft recovered to shore bases. *IWM (SP1156)*

was not long in coming. The claims of Col Billy Mitchell in the United States were put to a test (the conditions of which were subsequently disputed hotly) in 1921 when the old German battleship *Ostfriesland* was duly sunk by bombing, and many commentators were quick to announce the arrival of a new and overwhelming means of warfare at sea and the departure from the scene of the capital ship.

Naturally it was not quite so simple as that. The air weapon was still limited in many respects. Aircraft were slow, flimsy, of relatively short range; they carried few bombs, and these — despite the *Ostfriesland* experiment — were of doubtful lethality and accuracy of delivery; their navigation and communications were uncertain. Platforms for their deployment at sea, which had right from the start been regarded as a sound means of overcoming the range limitation, were still in the early stages of development.

Throughout the interwar period strenuous efforts were made by some sections of the naval services in all the major maritime powers to overcome the limitations and enhance the potency of the air weapon. From other, more conservative, arms of those services there was scepticism; their view was that the limitations were inherent, and that it was a waste of time and money to try to overcome them. The outcome in nearly all major navies was a compromise. Battleships remained, but air arms (both shore and ship-based) received a proportion of the available resources. That

Above:

The cruiser HMS *Sheffield*, one of the first British ships to be fitted with radar, a key component of World War 2 air defence.

IWM (A6163)

proportion was greatest in the navies of Japan and the United States, least in that of the German Reich. Britain occupied a halfway position where the air arm, particularly its seaborne element, was not neglected but where the battleship was still regarded as the primary naval unit.

Provision for defence against air attack reflected the preoccupations and — it may be — the prejudices of the navies concerned. Those navies that possessed aircraft carriers looked to carrier-borne fighters as the first and strongest element of their air defence; anti-aircraft guns, when mounted, were supplementary weapons, and indeed it was odd to see the main gun battery of the US carrier *Saratoga*, for example, consisting of low-angle 8in guns for surface engagement. In many British ships, anti-aircraft armament was confined to light-calibre weapons — the successors to the 'kind of pom-pom' mentioned by the 1913 piece quoted above. Of the destroyers in particular, only a small proportion in 1939 had main batteries capable of high-angle fire.

This under-provision — for, as the first two years of World War 2 proved, under-provision was what it turned out to be — was not just a matter of misallocation of resources. The learning curve was steep not only in the design and production of air defence weapons, but in their control systems. The difficulty of hitting — or exploding a shell in close proximity to — a target capable of relatively quick movement in three planes was considerable, particularly when the only means of rangefinding was visual and the only means of training and elevating the directing mechanisms, and the weapon mountings themselves, was by handwheel.

Thus, on both sides in those early years of World War 2, there were numerous instances of relatively small numbers of aircraft being able to press home attacks even against heavily-gunned ships. As aircraft operations grew less and less tentative, the shortcomings of air defence were further exposed; and with the audacious night attack by old, slow Swordfish aircraft on the Italian fleet in Taranto harbour on 12 November 1940, air power at sea came of age.

If there were those who still held that given sea room, ships could fight it out against unlimited numbers of aircraft for indefinite periods, then their convictions must have been sorely tried in 1941. In May of that year, two widely-spaced sets of events showed what might be expected. In the Atlantic the German battleship *Bismarck*, the most powerful ship afloat, was so severely wounded by air attack that, her mobility destroyed, she fell prey to British surface forces. In the Mediterranean, also in May, the British Fleet, endeavouring to support the defence of Crete and inhibit German seaborne invasion forces, suffered losses so crippling that even Adm Cunningham was forced to plead 'We cannot afford another such experience and retain sea control in the Eastern Mediterranean'. The casualties were extreme: two battleships and one aircraft carrier damaged, three cruisers and six destroyers sunk, six cruisers and seven destroyers damaged, all through air attack. Finally, still in 1941, came the sinking in the Far East of the *Prince of Wales* and *Repulse* by Japanese shore-based aircraft; clearly, armoured ships could not only be wounded, but actually sunk, in the open sea by air attack alone.

These events pointed to a single lesson, though the words in which it was expressed were sometimes more extreme than was justified. Perhaps the most accurate expression of it was naval historian Stephen Roskill's statement:

Right and below:
Royal Naval operations off Crete in 1941 demonstrated the vulnerability of ships to air attack. These pictures show a state-of-the-art gun's crew at action stations and Suda Bay under air attack after the Germans had established their air forces in Crete.
IWM (E1154 and E3140)

Right:
Many equipments did not stand up to prolonged action and sea conditions: it was necessary frequently to re-grease pompom ammunition.
IWM (A2089)

Left:
Two layers of air defence in World War 2: HMS *Ark Royal* operates aircraft in the background, with a cruiser's four-barrelled pompom in the foreground. *IWM (A2292)*

Below:
Before the US Navy gained the ascendancy, battles in the Pacific were desperate affairs: this is USS *Enterprise* under attack (and, with the accompanying battleship, offering stout defence) at the Battle of Santa Cruz.
IWM (EN21471)

'Control of surface waters by ships could not for long be maintained in the face of overwhelming air power. The answer to air power could only be air power . . .'

And, for oceanic campaigns such as were about to face the Allied forces — and particularly the United States forces — in the Pacific, that countervailing air power had to come very largely from the hangars and decks of specialised aircraft-carrying ships. Shore-based aircraft, even in that island-studded ocean, played a relatively insignificant part in the great sea battles that took place in 1942-44. These were a new sort of sea-fight: the main antagonists, the carrier forces, never set eyes on each other; aircraft were the 'main armament'; guns were used almost entirely for self-defence; totally new tactical factors, such as the state of the opponents' flight decks, came into play; damage-control vulnerabilities, particularly those concerning fire and high-octane vapour, were unusually critical. In these circumstances the older tactical prerequisites, information about enemy forces and communication with one's own, assumed even greater importance.

It was not surprising, therefore, that the main technical innovations and improvements during World War 2 concerned these very factors. First and foremost, command, control and information were revolutionised by the introduction of radar — both shipborne and, later, airborne — and of very high frequency voice radio. Carrier-borne aircraft, particularly the American types, improved enormously in ruggedness for deck operations, reliability and performance. Aircraft carrier damage control was greatly improved in the Allied fleets, though the British carriers with their armoured flight decks and closed hangars always had the edge here over their American counterparts. Finally, ships' gun systems for self-defence against air attack proliferated. Towards the end of the war, surface units positively bristled with anti-aircraft armament, mostly of the close-range sort, the mountings more and more generally operated by remote power control and directed by radar.

The last great carrier battle of World War 2, the Battle of the Philippine Sea in 1944, demonstrated American superiority in nearly all these factors. To be sure they had greater numbers, but their technical edge was also considerable. Aided by good work from their submarines early in the action, they completely broke the power of the Japanese in the air in what became known as the 'Great Marianas Turkey Shoot' before turning to the offensive themselves. The Battle of the Philippine Sea is instructive not only because it shows how comprehensive, well-directed and well-timed air defence can lead to success in the business of attacking the enemy, but because it has burnt itself into the consciousness of the United States Navy: not perhaps so deeply as Trafalgar into the Royal Navy of the 19th and early 20th centuries, but significantly in a way that it will be necessary for this book, later on, to discuss.

Meanwhile on the other side of the world, the Allied experience — in this case preponderantly British — was somewhat different. The menace to the Atlantic convoys was very largely from under the water, but long-range German aircraft operating from bases in occupied France were a threat and needed checking by local air cover, which was provided by improvised or specially-constructed light aircraft carriers. In more restricted waters, different measures were needed; shore-based air power was much more generally employed to give the first line of defence against air attack, and for the invasion of Normandy in June 1944 that rare phenomenon, total control of the air space over a seaborne force, was achieved by the land-based Allied air forces. A year before, during the Salerno landings, it had not been so clear-cut; and several warships escorting the

Below:
Post-World War 2 technology has greatly increased the power of the offensive. The Soviet aircraft carrier *Novorosiisk* has both VSTOL aircraft and long-range surface-to-surface missiles. *Crown Copyright*

invasion force were damaged by a new kind of airborne weapon, a glider bomb controlled in flight to its target by its parent aircraft. It was a portent of many things.

For, great as had been the technical advances just before and during World War 2, there were other impending developments that made the subsequent period of uneasy peace and small wars — which has gone on until the present day — one of almost equally rapid change.

First, aircraft and airborne vehicles were about to become much faster with the advent of jet engines, rocket motors and their variants. Second, the extremely rapid advance of technology in the fields of radar, electronic processing, gyro-controlled stabilisation and inertial navigation systems, servo mechanisms, materials and fuel made it possible for weapons as well as their carriers to be much more precisely controllable, accurate and autonomous. Finally, the development of nuclear weapons to the point where they could be used not only as bombs but as the

Above:
Modern warships of medium size are thin-skinned and may be vulnerable to a single hit. HMS *Ardent* was lost to air attack off the Falklands in 1982. *Crown Copyright*

Left:
Up-to-date technology can help the defence too: this is part of the operations room of a modern frigate. *Ferranti Computer Systems Ltd*

warheads of missiles gave an added dimension to the problems of air defence at sea.

So severe did those problems seem to some commentators that they despaired of the utility of surface naval forces. Particularly in the medium powers, many pundits and politicians argued that 'navies must go underwater in order to survive', and even in the superpowers some opinions could be heard to suggest that surface ships were 'sitting ducks' or 'coffins for the crew'. Such views were, of course, reinforced by the parallel strides being made by the other great threat to surface ships, the submarine arm.

However, these views have not prevailed for a very simple reason. Sea power is the ability to use the sea; the use of the sea is overwhelmingly the use of the sea's surface; and to secure the use of the surface, fighting units on the surface — as well as all the other sea-users, merchant ships, transports, fishers and oil rigs — are required. There are no easy options here.

In consequence, the provision of air defence for surface units has been a major preoccupation, and a 'real big spender', for maritime powers world-wide. Technological change has much increased the options and improved the possibilities; in many cases the very same advances that sharpened the threat have also made available the means for countering it — but at a price which often does not look a very attractive trade-off.

For, when all is said and done, surface ships are large and complex units, capital and manpower-intensive; usually these days they have thin scantlings; they are almost always unarmoured. Packed with power sources and fuel, they are as prone to flash and fire as ever they were; and even though the most lethal of all conventional airborne weapons, the torpedo, is no longer used for anti-ship work, a single missile or bomb can incapacitate or — as was shown in the Falklands campaign — even destroy.

Thus, keeping the sea in the face of a modern air threat will be a difficult and expensive business, calling on a wide range of technologies and tactical skills. In many cases, particularly for medium powers, resource constraints will make it impossible to seek perfection. The use of the sea will turn, as so often it turned in the past, into an attrition operation; casualties will have to be expected. It may not be possible to contemplate certain operations because of an overwhelming air threat. The superpowers, on the other hand, may aspire to a more comprehensive and leak-proof air defence for their maritime forces. But whoever is conducting this technically-driven, swiftly-moving, immensely complex task of air defence at sea, by whatever means and with whatever resources, they will be doing it for what to them seems sufficient reason: it has to be done.

CHAPTER 1

The Threat

There are two good reasons for addressing, in the bulk of this chapter, the threat to Western surface units from Soviet maritime forces. First, this is after all a book written from a Western standpoint; and though at a later stage of the chapter some attempt will be made to look at the air defence problem through Soviet eyes, the main thrust of threat discussion must reflect what is perceived by the West. As was explained in an earlier book (*Anti-Submarine Warfare*, Ian Allan Ltd, 1984), so long as the Soviet Union maintains the Marxist dynamic, the perception of the Soviet Union as a threat must be a realistic one. So long as deterrence — which in large part depends on the maintenance of effective defence including air defence — goes on working, the threat will remain just a threat, and we'll all drink to that.

Secondly, the Soviet air threat is the most comprehensive of all those that are wielded by the world's maritime forces, spanning as it does so many modes of warhead delivered by so many types and ranges of vehicle from such a diversity of platforms. It is therefore a suitable exemplar, in planning jargon. Published material on most of the elements of this threat is not so comprehensive, of course, and it is sometimes necessary to read across from Western state-of-the-art technology to get an idea how the Soviet equipments might operate.

That will be addressed also; and this in its turn will serve to consider the threats that might be faced in those situations where the superpowers are not involved. It should be remembered that in the past two decades these have included not only the Falklands campaign, but the Indo-Pakistan War of 1971 and the Yom Kippur War of 1973, as well as several other maritime operations where air defence no doubt played a prominent part in the plans even though it was not required in practice.

Soviet Objectives

The defence of the homeland is an objective deeply rooted not only in Russian character but in Soviet ideology; for, without a secure base, the advance of Communism could not flourish and the correlation of forces would be at least dubious, possibly adverse. Thus, when in the late 1940s the Soviet Union perceived a threat building from the United States aircraft carrier forces, measures were put in hand to counter it by every possible means. A great many of these measures involved air vehicles, and in modified form they continue to this day.

However, the development of Soviet naval thought under Admiral of the Fleet of the Soviet Union Sergei Gorshkov, Commander-in-Chief from 1956 to 1985, was not confined to the defence of the homeland against American carrier battle groups, nor to the use of the Navy in support of flanking moves by the Army in the old Russian tradition. Gorshkov was insistent that 'fleet against shore' — his shorthand for the capacity to conduct strategic nuclear bombardment against the vitals of enemy territory at any stage of a conflict — was a primary role for the Soviet Navy, and one that would demand a heavy allocation of naval resources not only for the main platforms for such operations, the nuclear-powered strategic ballistic missile submarines (SSBNs), but for forces that could protect and safeguard them.

In consequence, and aided by the ever-increasing range of ballistic missiles, there was evolved a Soviet strategy of SSBN 'bastions' in the Arctic and sub-Arctic waters relatively close to the Soviet Union. These are to be held inviolable, a key area before, during and at the end of any war. Thus, in both Atlantic and Pacific theatres, there is a considerable extension seaward of areas regarded as vital by the Soviet Union; and these too will be held to require a powerful threat to counter any Western forces that venture there. The fact that Gorshkov's successor, Adm Chernavin, is a submariner suggests that there will be little change in that particular objective.

That is certainly not the end of the tale. The Soviet Navy has since the early 1960s maintained a presence in the Mediterranean and the Indian Ocean, and since the mid-1970s in the China Sea.

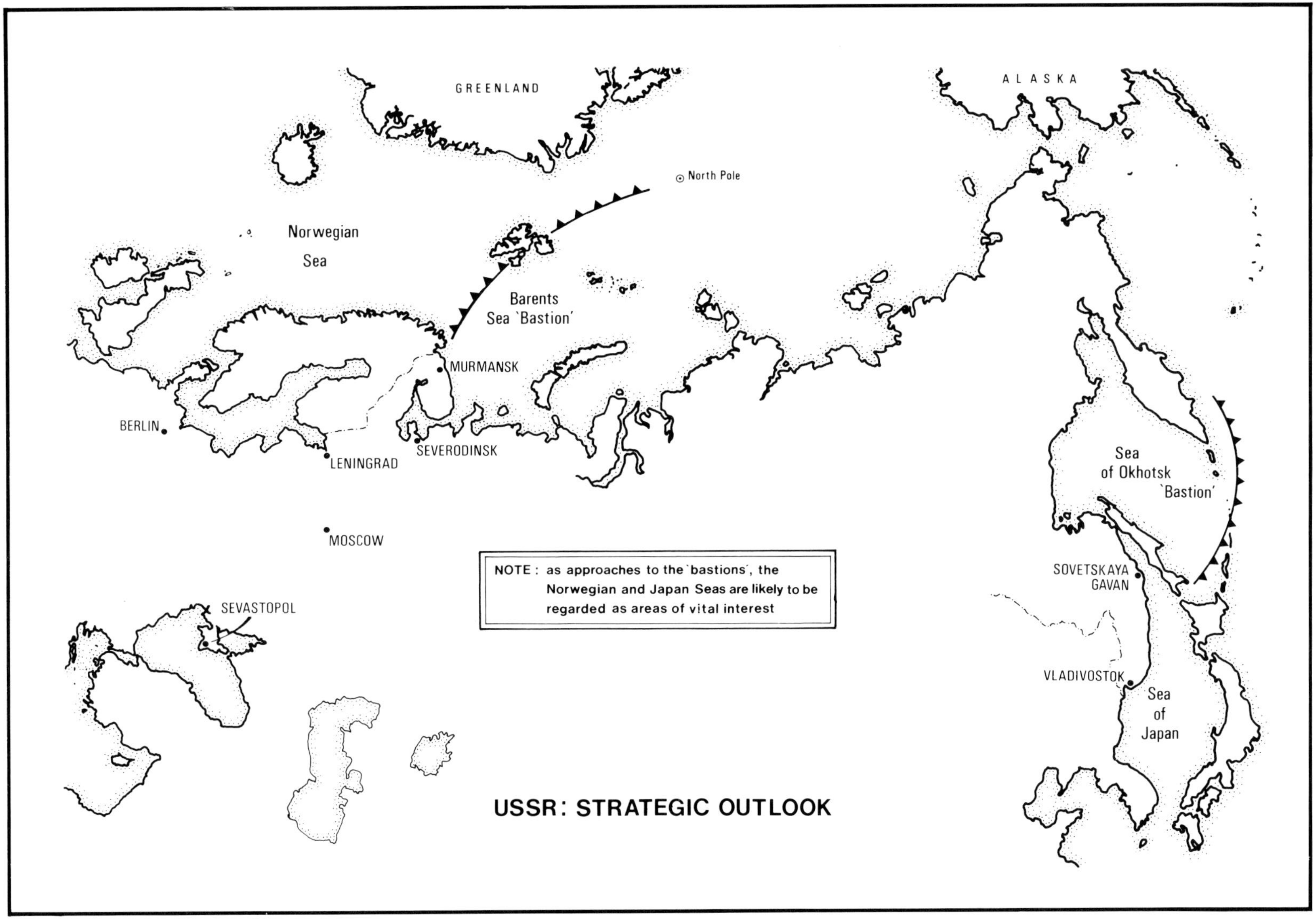

USSR: STRATEGIC OUTLOOK

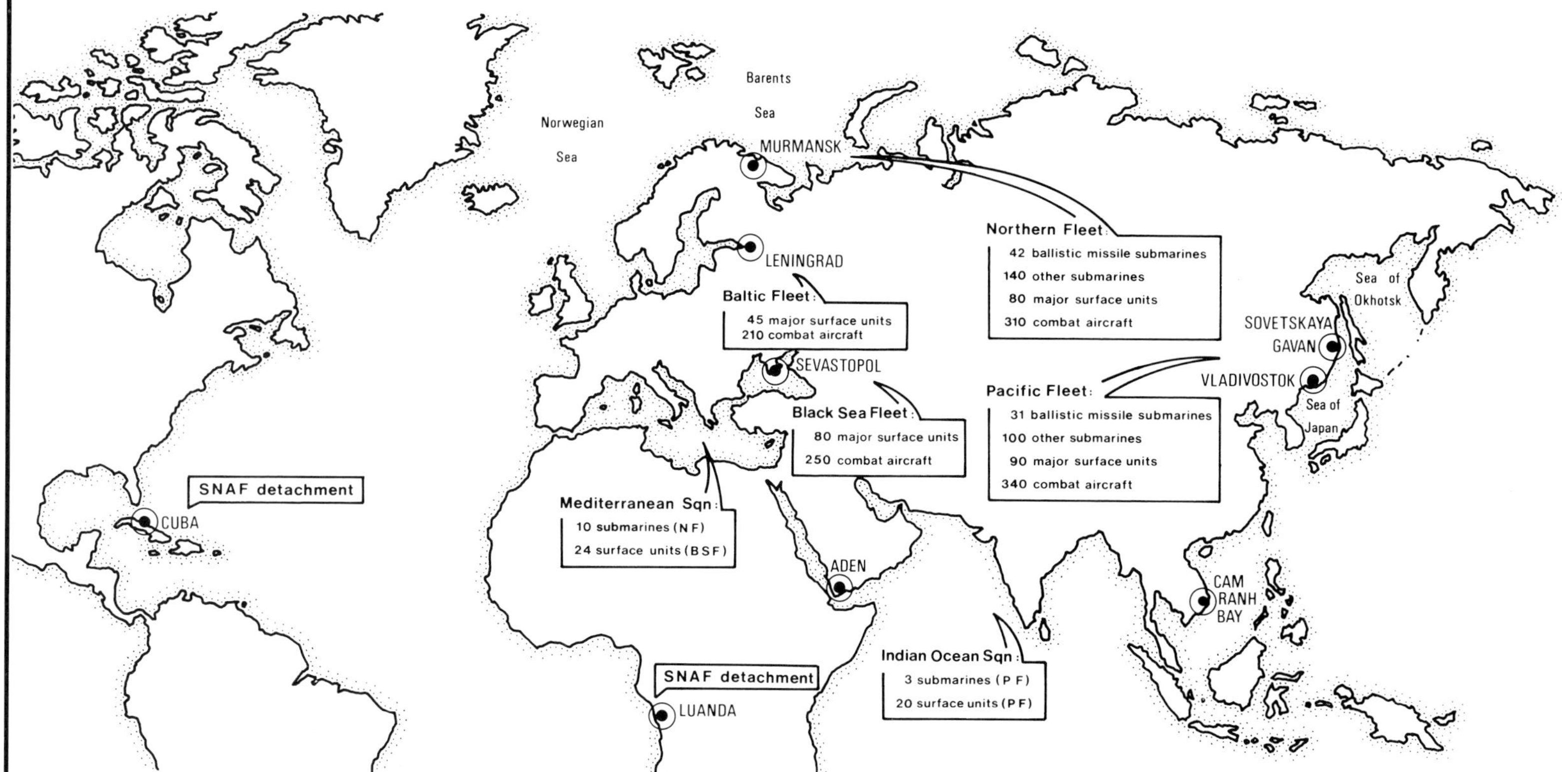

PRINCIPAL SOVIET MARITIME POWER CENTRES

Its forces in these areas are often beyond shore-based air cover, so they have severe air defence problems of their own, but they do also pose an air threat, in varying dimensions according to the forces deployed, to any opponent. In so doing they do of course very greatly extend the reach of the Soviet surface navy, which in the 1950s was confined to a few hundred miles of the Russian mainland. How such forces would be used, and in what way they would be expected to further Soviet objectives, is a matter still for speculation. You can roll your own scenarios; the opportunity is there and the threat cannot be ignored.

Intelligence, Reconnaissance and Surveillance

Accurate and timely knowledge of the position and movement of all relevant opposing forces is necessary for effective air attack. Moreover, knowledge of the characteristics of those forces is also highly desirable and may be vital for success. The Soviet Union thus sees a need to build up a comprehensive data base on Western maritime forces, to monitor their movements in normal conditions and to maintain an accurate plot on them, sufficient to provide target data, in a period of tension or in areas sensitive to the USSR when the 'first salvo' made famous by Gorshkov's writings, and now enshrined in Soviet naval tactical doctrine, might be called for.

The Soviet machinery for providing this intelligence, reconnaissance and surveillance is impressive in scale and, by all accounts, in performance. It employs vehicles in outer space, in the air, on and beneath the sea, as well as shore stations and a comprehensive data exchange, command and control network. In Western terminology it goes by the name of Soviet Ocean Surveillance System (SOSS).

The outer-space layer of the system consists of a variety of satellites. They can be divided into two categories. First come time-late systems, which provide additions to the data base, though — if delays are not too great — they may also supply indicators of Western intentions. These are photographic, electronic intelligence (ELINT) and direction-finding systems. The other category can properly be called real-time: satellites which provide instantaneous information on targets within their field of view. Generally these are radar satellites, though in certain circumstances direction-finding satellites can be considered real-time systems.

There are perhaps two major points to be made about the space component of SOSS. First, it is one of those complexes where the whole is greater than the sum of its parts. Putting together the time-late and real-time elements, if this can be done with certainty, delivers a picture that holds. But a simple read-out from a radar satellite carries all kinds of possibilities of misreading and wrong interpretation, particularly if cover of the sea area concerned is only intermittent (because of the difficulty of putting such satellites in geosynchronous orbit). Second, it is extremely expensive, all the more so if the redundancy necessary to build a complete picture is provided — for example, if radar satellites have to be flown in

Below:
The Tu-95 'Bear-D', seen here accompanied by an F-14 Tomcat from USS *Nimitz* (background), is still an important component of the Soviet Ocean Surveillance System. *US Navy Official*

sufficient numbers to give comprehensive cover of certain sea areas.

Thus, although such cover from space may be a Soviet aspiration, they are certainly not going to rely on it exclusively for their surveillance needs. In any event they would consider it grossly imprudent to do so because of the risk of anti-satellite weapons being used against them in the early stages of a conflict.

The next element of SOSS is, not unexpectedly, long-range reconnaissance aircraft. The main Soviet vehicle in this field is a big, solid middle-aged turboprop aircraft with the NATO nickname 'Bear'. This has the virtues of long endurance on task, reliability and adaptability to new equipments, and it is little wonder that the Soviet Navy has kept the aircraft in its inventory for 30 years and is expected to persist with the design, in an evolving form, for the rest of the century. The version with which this book is chiefly concerned is the 'Bear-D', an aircraft whose mission is above-water surveillance, reconnaissance and provision of target data. Its chief sensor is a search radar but it has a full suite of passive listening devices that can detect ships' electronic emissions. Its communications fit is comprehensive and includes a video data link to missile-armed units at sea and, possibly, in the air also. 'Bear-Ds' have for over two decades been a familiar sight to Western units, particularly US carrier battle groups (CVBGs) on passage or exercises. In recent years, through their basing at Cuba, Conakry (replaced in 1977 by Luanda) and Cam Ranh Bay, they have greatly extended their area of potential cover.

Some other Soviet aircraft, both shore and ship-based, can be used for reconnaissance and targetting roles, though they would not be fully integrated with SOSS. Such aircraft are the shore-based 'Backfire', 'Blinder', 'Badger' and (on an opportunity basis) the ASW 'Bear-F', 'Mail' and 'May'; and the ship-based vertical take-off 'Forger' and the 'Hormone' and 'Helix' helicopters. Some of these will re-enter the discussion when missile systems are being considered.

But there are some elements of SOSS that are neither air nor space-based. First are the shadowing units, generally surface ships, that attach themselves customarily to US CVBGs in the Mediterranean and in any other area of interest to the Soviet Union. They are often nicknamed tattletales, which fairly precisely describes their mission. By providing a constant input to SOSS, as well as a mass of data for future analysis, they fulfil both the time-late and real-time function. Certainly their life when the shooting started would be strictly limited, but if they had been instrumental in achieving an accurate and weighty 'first salvo' from a variety of other units, they might feel their job had been well enough done. Submarine tattletales can fulfil the same functions, less comprehensively perhaps but with marginally better chances of survival if the balloon goes up.

There are also the intelligence-gathering ships, usually in the shape of trawlers, which generally occupy geographical stations in the approaches to areas of Soviet interest. These have a more general electronic listening function and their real-time contribution to SOSS is probably fairly limited; it seems doubtful that their prime function is to provide targetting data, though they might be expected to do so if opportunity offered.

Finally, the extensive network of Soviet ground listening and direction-finding stations is an essential element of SOSS and must not be forgotten. Probably the oldest part of the system but constantly modernised and updated, it is a powerful means of providing near-real-time information on any unit transmitting on high or medium frequencies, and of gathering a mass of electronic information for subsequent analysis and addition to the data base.

Weapon Systems

While the Soviet Union's weapon procurement has often resulted in the adaptation of one weapon for use from a variety of platforms, sometimes widely diverse platforms, it is convenient generally to consider the weapon and platform as a complete system, and that is the method that will be adopted in the following discussion.

Air-based Systems

The free-fall bomb is still in the Soviet weapon inventory, but is not widely employed. It is the chief weapon of only one type of shore-based naval aircraft, the ageing 'Blinder' which is deployed only in the Baltic. It could also be used by first-generation carrier aircraft, but even these aircraft have other attack systems and a high degree of training in bombing is unlikely. The Soviet Naval Air Force has concentrated for many years on missile solutions.

It was soon into this field. As early as the 1950s two types of long-range, aircraft-borne missile were deployed by the Soviet Union. One was nuclear-headed and of several hundred miles range; called the 'Kangaroo' in NATO parlance, it was carried by the 'Bear' in its 'B' or 'C' variants and was generally subordinated to the Soviet Long-range Air Force (LRAF). It is not clear whether the 'Kangaroo' was ever considered for tactical use against ship targets, but two things are certain: first, it is still in service; and second, a long-range nuclear shot 'into the brown' of any concentration of ships could cause havoc. So it cannot be ignored.

The other 1950s air-to-surface missile, the

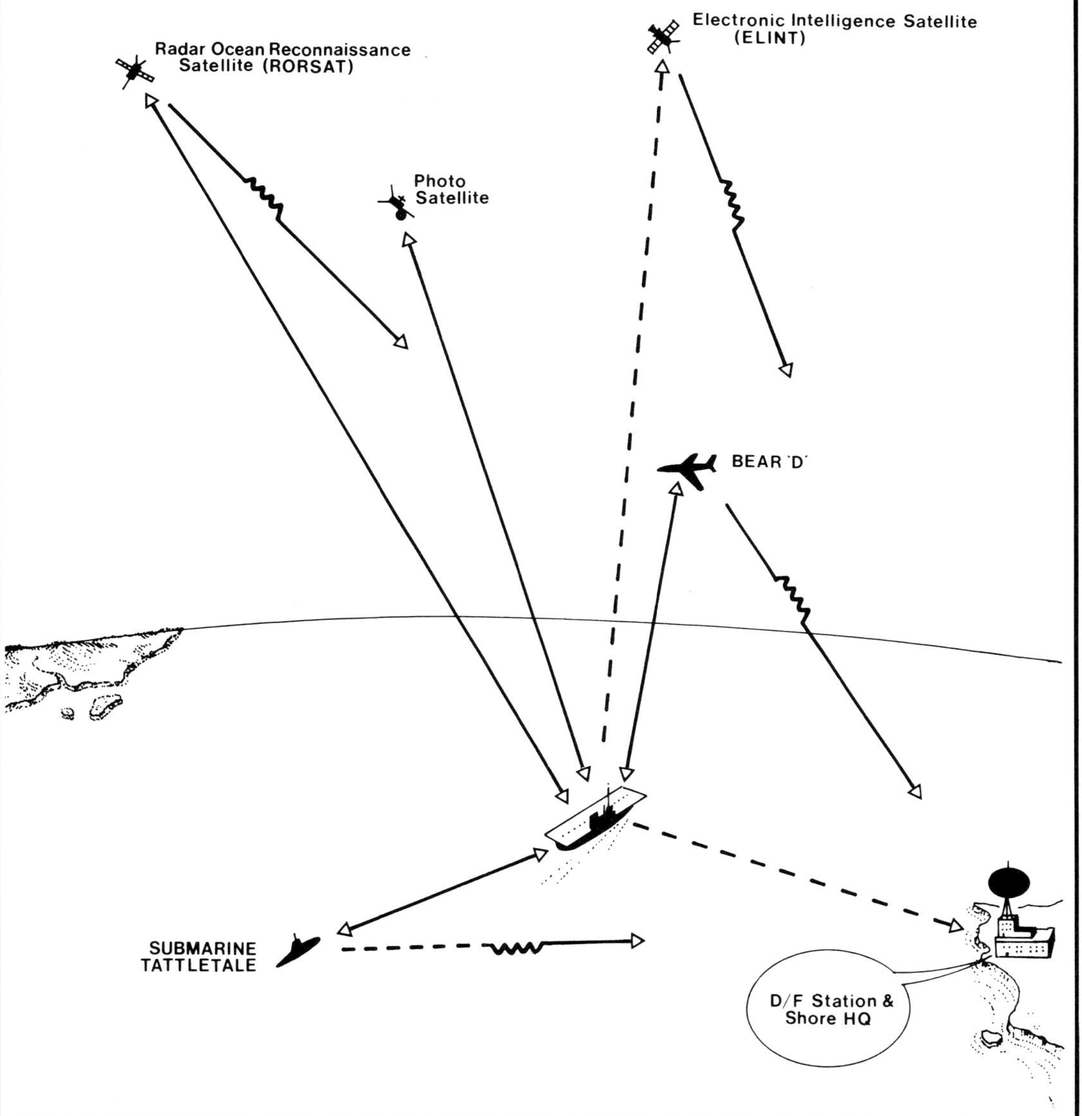

TARGETING OF A CARRIER FORCE BY SOVIET OCEAN SURVEILLANCE SYSTEM (SOSS)

- FIRM CONTACT BY ACTIVE SENSOR, TRAIL OR OPTICAL MEANS (solid double-headed arrow)
- CONTACT ONLY IF TARGET TRANSMITS (dashed arrow)
- REAL TIME CONTACT WITH HQ (solid arrow with wavy section)
- OCCASIONAL REAL TIME CONTACT WITH HQ (dashed arrow with wavy section)

NOTE: photographic satellites are essentially long-term intelligence gathering systems. Photographic information on a ship at sea would be fortuitous and not delivered at once.

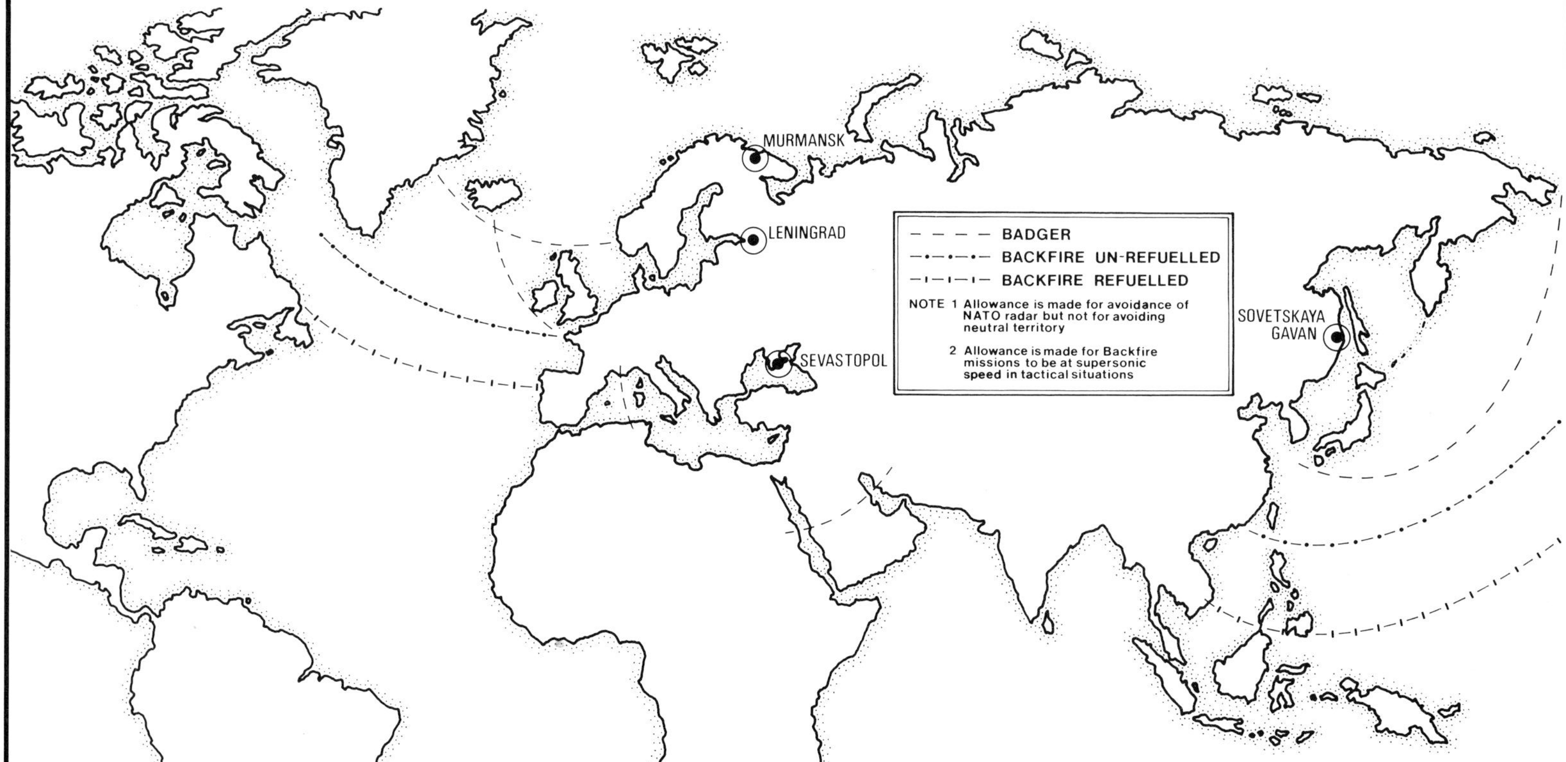

SOVIET MISSILE-ARMED AIRCRAFT COVERAGE FROM MAIN BASES

‘Kennel’, is long gone, but it was the forerunner of several generations of weapon, most of which survive. Nearly all are, astonishingly, carried by the same aircraft that pioneered the ‘Kennel’ so long ago — the ‘Badger’, like the ‘Bear’, a product of the Tupolev stable and one of the long-running success stories of Soviet Naval Aviation. The ‘Badger’ is a twin-engined, shore-based subsonic aircraft whose endurance allows it to reach down to the Iceland-Faeroes Gap from the Kola bases or into the Sea of Japan from Sovetskaya Gavan. Flight refuelling gives it extra range potential. The ‘C’ variant can carry one old, fairly slow ‘Kipper’ missile; or two more modern, faster ‘Kingfish’ missiles. Both have a range, after leaving the parent aircraft, of over 100 miles — the ‘Kingfish’, by report, considerably more. It is a prudent assumption that the ‘Kipper’ will be seen less and less and the ‘Kingfish’ more and more. The ‘G’ variant of the ‘Badger’ originally carried the ‘Kelt’ missile of just 100 miles range, but can now fit the ‘Kingfish’ instead.

One current aircraft/missile system is not in the ‘Badger’ line of succession. This is the potent ‘Backfire’, a shore-based, variable-geometry supersonic aircraft introduced in the mid-1970s, which gives a further element of urgency to the air defence problem. Moreover, it increases its scope; with reasonable fuel conservation the ‘Backfire’, even without inflight refuelling, considerably out-ranges the ‘Badger’. The missile carried by the ‘Backfire’, which is nicknamed the ‘Kitchen’, is also fast and of long range — nearer 200 than 100 miles.

Recently, commentators have noted in the naval air defence context the appearance of the long-range supersonic ‘Blackjack’ bomber. This is still in the development stage and its primary role is probably not maritime. But if Soviet precedent is anything to go by, a naval version, almost certainly missile-armed, will eventually appear. This will increase the range and delivery speed of the threat.

All the missiles described above probably have the same kind of guidance system: that is to say, they do the first part of their independent flight under inertial guidance and the final homing is by active radar in the head of the missile. However, there is ample room within this basic configuration for variations such as infra-red, home-on-radar or home-on-jam characteristics, and for a variety of flight patterns as the missile approaches its ship target: descending to sea-skimming height, bunting to achieve a steep terminal dive or just coming straight on in.

There is also room for variation in the nature of the warheads. Apart from the ‘Kangaroo’, which

Below:
Soviet carrier-borne fixed-wing aircraft are so far limited to the VSTOL Yak-38 ‘Forger’, of which two are seen ranged aft in the carrier *Kiev*. *Crown Copyright*

makes little sense unless used in the nuclear mode, the missiles described can have either conventional or nuclear warheads; the conventional range is between 1,000lb and 2,200lb and the nuclear would perhaps reach up to 200 kilotons. These are heavy charges by Western standards, and even in the conventional mode one hit is enough to do grave damage to a medium-sized surface unit.

Soviet carrier-borne aircraft have been left out of the account above, and for completeness it is necessary to make a note about them here. So far the only fixed-wing carrier aircraft is the VSTOL 'Forger', a severely limited aeroplane compared with the Sea Harrier. Lacking radar, it is suitable for day operations only, and its radius of action is extremely short. It can deliver either a relatively unsophisticated strike with short-range air-to-surface missiles (maximum range six miles) or with free-fall bombs. However, the Soviet carrier-based potential should not be dismissed indefinitely. The first large fixed-wing carrier, of about 70,000 tons and nuclear-powered, was launched at Nikolaev in 1985. It will be capable of operating much more sophisticated aircraft than the 'Forger', though reports suggest that, rather than specially designed carrier aircraft, these will be marinised versions of tried types such as the 'Flogger', 'Fencer', 'Flanker' or 'Fulcrum'. These are capable of all-weather operation and may well be able to carry weapons of much more range and potency than the 'Forger's' load.

Ship-based Systems

If the Soviet Union was among the leaders in fitting air-based anti-ship missile systems, in the field of operational deployment of such systems from ships it was a pioneer. The development was strongly driven by the 1950s perception of a carrier air threat to the homeland: a threat that was considered to demand an all-arms counter. Thus, although the vulnerability of surface ships as missile platforms was acknowledged (sometimes in brutal terms, as in Kruschev's famous 'coffins-for-the-crew' remark) they were viewed as a necessary addition to the counter-carrier armoury.

The first system appeared in the late 1950s, fitted in a class of large destroyer called the 'Krupny'. The missile was a typical first-generation device, a crude little subsonic aircraft nicknamed 'Strela' (if memory serves), with an active radar homing head and inertial mid-course guidance. The standard warhead was nuclear. The writer took part on the Orange side in the first British exercise concerning these things, in 1961. Aided by the US Air Force for target indication, and using all the manoeuvring subterfuges dear to British destroyer forces, we managed to close undetected to within the 100-mile firing range and got off our 'missiles' in a strike judged to be successful. It was apparent at the subsequent wash-up that the carrier force had not taken the threat very seriously, thinking that by the usual reconnaissance, probe and strike they could quite easily eliminate the 'Krupnys' before they closed to firing range. Their resources were also stretched by a concurrent air threat. The lesson was clear: the ship-based threat could not be ignored.

Soon it became more potent and of longer range. In the early 1960s there appeared the 'Kynda' class cruisers with two monstrous sets of missile tubes housing altogether eight 'Shaddock' missiles. These missiles were — and are, for in both their original and subsequently modified form they are still operational — large birds, over 30ft long, with a speed in the high subsonic area and a range of up to 250 miles. In order to use this range effectively it is necessary to give mid-course guidance to these missiles; this is done by a missile

SOVIET AIR-TO-SURFACE MISSILES

In service	*Designation*	*Length (ft)*	*Launch weight (lb)*	*Warhead (lb/kt)*	*Speed (Mach)*	*Range (nm)*	*Guidance*	*Launch vehicle and remarks*
1961	AS-2 'Kipper'	30	9,270	2,200/200	Subsonic	110	Autopilot, active radar homing	'Badger-C' and 'G'
1961	AS-3 'Kangaroo'	45	?	Nuclear up to 800kt	Subsonic	350	Inertial	'Bear'
1967	AS-4 'Kitchen'	37	14,350	2,200/200	3.5	150	Inertial, active radar homing	'Backfire'
1965	AS-5 'Kelt'	28	10,375	2,200/200	Supersonic	100	Inertial, active radar homing	'Badger-C' and 'G'
1970	AS-6 'Kingfish'	34	10,800	1,100/200	2.5	200	Active radar	'Badger-C' and 'G'. High trajectory and terminal dive
1980	AS-7 'Kerry'	?	770	200lb	1.0	6	Active radar	'Forger'. Some authorities suggest radio command guidance
1980	AS-9	?	?	350lb	3.0	60	Home-on-radar	'Badger', 'Backfire'

Note: A proportion of all missiles may be fitted with alternative infra-red homing.

Above:
A 'Kynda' class cruiser, an early example of a ship designed primarily to deploy long-range surface-to-surface missiles.

Below right:
Carrying the SS-N-19, one of the newest types of long-range anti-ship missile, is the 'Kirov' class battlecruiser *Kirov*. *Crown Copyright*

guidance radar in the parent unit, though in turn this unit may use target information from other sources. Terminal homing is generally by active radar in the nose of the missile, though infra-red may well be incorporated. The warhead is of 2,200lb and can be either conventional or nuclear.

At the same time as the 'Shaddock', a much shorter-range missile, handy for fitting in fast patrol craft, made its appearance. This was the 'Styx'. Like its contemporaries it was liquid-fuelled, subsonic, had active radar homing and was not very sophisticated. Unlike them, it did not rely on extraneously gathered target data, since its range of 16 miles was within the field of radar view of its parent unit. The 'Styx' became very widely distributed during the 1960s, and not only round the coasts of the Soviet Union; it was a very good seller among Soviet arms-sales clients and, in 1967, made history when an Egyptian 'Komar' class patrol boat in Port Said fired several missiles at the Israeli destroyer *Eilat*, some 12 miles off the coast, and sank it. This was the first ship-to-ship missile casualty. It led to much sensational treatment in the press, which seemed to think that their ignorance of the threat was shared by the world's navies. That was not so; but it is fair to say that if there had been complacency in some quarters before, there was much less afterwards.

'Shaddock' and 'Styx' remain in service. Improved and updated versions, the 'Sandbox' and 'C' variant of the 'Styx', were introduced in the 1970s and fitted in the more modern classes of Soviet surface ship from the 'Nanuchka' class corvettes right up to and including the 'Kiev' class carriers. These missiles, while embodying much re-engineering and more advanced technology that almost certainly improved flight profiles, reliability, maintenance and ease of handling, were also of increased range; and, while sharpening the problems of the opposition, this led to operational complications for the firing unit and its supporting forces. In particular, once the 'Styx' in its 'C' variant was given an over-the-horizon range, it required a new kind of target indication to make full use of that range. In the larger classes of ship (and the 'Styx-C' has gone into ships of destroyer

size) this may be provided by a helicopter, but in the smaller units it requires either shore-based aircraft or skilfully-placed surface units.

The same problems associated with extended range apply to the newest of the Soviet ship-based missile systems. These are fitted in the nuclear-powered 'Kirov' class battlecruisers — which have a supersonic, very long-range missile as yet unendowed with a NATO nickname — and the 'Sovremennyy' class large destroyers — these have a very fast missile of about 70 miles range with a solid-fuel motor, likewise so far given only a number in NATO designation. The second of these missiles has also been installed experimentally in a corvette of the 'Tarantul-II' class.

This new generation of ship-based anti-ship missiles almost certainly incorporates the sea-skimming characteristic that presents, as will be made clear in the next chapter, one of the most difficult problems of air defence at sea. Previously, this had been apparent only in the 'Sandbox' missile and then only in the very last phases of its trajectory. Now, it appears to be of much more general application. When that is taken with the very much higher speeds of the new-generation missiles, they can be seen as formidable threats once launched.

Submarine-based Systems

In the late 1950s some weirdly-fitted submarines began to appear at the Leningrad Navy Days. These were variants of the widely-distributed 'Whiskey' class, mounting, as quickly became clear, very large missiles in tubes external to the pressure hull. In some variants the tubes could be elevated, in others they were incorporated in a kind of bloated fin.

Unpretty sights as they were, these submarines had some propaganda value as a signal of the Soviet intention to take large missiles to sea in submarine platforms. Production classes were not slow to follow, and the sequence has continued to the present day. The first two such classes were the nuclear-powered 'Echo' class and the conventionally-powered 'Juliett' class. Both carried paired launchers; the 'Julietts' had two sets, the earlier 'Echos' three and the later 'Echos' four. The missiles turned out to be none other than the 'Shaddock' already (though no more than a year

Below:
An 'Echo-II' class submarine mounting eight surface-to-surface anti-ship missiles.

before) seen deployed in the 'Kynda' class cruisers. This gave the Soviet Navy a new diversity in its anti-surface ship armoury: one which, because of the potential concealment of its platform, presented a worrying problem.

However, the system was not without its limitations from the Soviet point of view. First, submerged launch was not possible. The submarine had to expose itself for a considerable time — perhaps several minutes — before being ready to fire. Second, it had to get its target information from somewhere; and communications never were a submarine's strong point. Finally, the missile would probably need corrective guidance in mid-course, and although the submarines were fitted with a radar for this purpose, that meant that they had to remain exposed at a time when they would dearly have liked to slip away. Again, submarine radars in the early 1960s were not noted for their reliability.

It was no surprise, therefore, that the next Soviet submarine mounting anti-ship missiles was an altogether more autonomous affair. This was the 'Charlie' class, a nuclear-powered boat of medium size with a second-generation power plant and improved hull form. It carries eight missiles in purpose-built tubes, which are fitted in addition to the usual torpedo tubes. The missiles have two principal operational characteristics: they can be launched from beneath the surface, and they are of relatively short range. This means that the submarine can fire on its own acoustically acquired information while keeping its own exposure to the minimum. Naturally there are tactical risks involved; acoustic information is not so certain nor so accurate as that which is gathered by visual or radar means, and even a submerged approach to

Top:
A 'Charlie' class submarine, incorporating the first Soviet submerged-launch anti-ship missiles.

Above:
Latest and largest of Soviet guided-missile submarines is the 'Oscar' class.

within 25 miles of a force puts the submarine at some hazard. But ask any submariner whether he would rather have the 'Echo'/'Juliett' or the 'Charlie' system to operate, and he will almost invariably choose the latter.

Perhaps it was a desire to get the best of both worlds that led the Soviet Navy to embark on the latest addition to its armoury of anti-ship-missile armed submarines, the monstrous 'Oscar' class. This vessel is of nearly 15,000 tons submerged displacement, nuclear-powered of course, and mounts 24 missiles of very long range, thought to be the same type as those carried by the 'Kirov' class battlecruisers. This would give them supersonic speed and a variety of flight profiles with, by some reports, programmed simultaneous arrival at the predicted position of the target. They can be launched when the submarine is submerged. The arrangements for their mid-course guidance in such circumstances are uncertain, and might indeed still be under development; reports suggest that satellites may be involved. The 'Oscar' class became operational only in 1983 and if previous Soviet practice is anything to go by, it has a long life ahead of it.

Future Developments

Even though it does hold on to its old gear for long periods, the Soviet Navy's track record suggests that it regularly introduces new and improved systems, sometimes evolutionary, sometimes radically different. The next development is expected to be a cruise missile of about 1,000 miles range, similar in concept to the anti-ship version of the US Tomahawk. Soviet technology is probably quite capable of reproducing the low-altitude characteristics of such a weapon, propulsion is no problem, and a capacity for a variety of terminal manoeuvres to confuse and make life difficult for defences is well within the Soviets' grasp. More doubtful is their capacity to make the missile intrinsically difficult to detect by radar (stealth technology depends on very high engineering as well as design standards) and to target it accurately in confused tactical situations. But such a development, even if only for the additional protection that its extended range gives to its air, surface or submarine platform, is an added preoccupation for the West.

Target Acquisition and Updating Problems

Several times in the above catalogue it has been suggested that one of the principal problems facing the Soviets is reconciling the generally long range of their missiles with the uncomfortable facts of geography, geometry and physics: namely that the earth is curved, visual and radar lines of sight are straight, and consequently the dog often cannot see the rabbit. This problem is least acute when the weapon platform is airborne; a 'Badger' releasing a missile of 100 miles range at 15,000ft can easily acquire its target with its own radar. But for ship and submarine platforms firing 'Shaddock' or 'Sandbox' missiles, for example, at full range, target acquisition must be done through some other agency. Undoubtedly this is one of the principal reasons for the complexity and comprehensiveness of SOSS, and it may be that by now the Soviet High Command is reasonably sure that enough information will be available, certainly in the stages leading up to conflict, to establish a full picture on the principal targets and pass that information to all units that require firing data. That will have been no light task; collection, collation, evaluation and distribution of the data is a process that can break down at any point, and it is almost certain that in parts at any rate it will so break down. What Soviet leaders no doubt hope is that, given plenty of system redundancy, enough training (as in the 'Okean' series of exercises) and a multiplicity of firing units, the breakdowns will be fragmentary and insufficient to matter.

There is also, however, the problem of updating target information once the long-range missiles have been launched. As an illustration, a 300-mile range missile travelling at 600mph takes half an hour to reach its target if fired at extreme range; during that time the target may well have travelled 12 or 15 miles, enough to take it out of the arc of search of the missile's radar homing head. Updated target data is needed to provide the necessary mid-course guidance and, as has been shown, this cannot be provided by the firing unit. Either, therefore, the unit providing the updated information must pass instructions based upon it direct to the missile, or it must pass it to the firing unit which then passes instructions to the missile. The first method entails the swift establishment of a link to the missile, once it is fired, by a unit which has never been in contact with it before; the second entails establishment of communication quickly between the firing unit and the reconnaissance unit, and a lot of double processing thereafter. Neither is at all easy to manage in a tactical situation at sea.

Soviet Tactics

Soviet tactical doctrine distinguishes sharply between the offensive and the defensive, and in general greatly favours the former. Offensive action is, on land, to be concentrated against limited sections of the enemy front and then exploited swiftly. At sea this principle appears to have been modified: concentration is indeed to be the governing factor, but it is concentration in time as well as place; priority targets are to be the enemy's main or most immediately dangerous forces in the 'fleet against shore' context; and

SOVIET SURFACE- AND SUBSURFACE-TO-SURFACE MISSILES

In service	*Designation*	*Length (ft)*	*Launch weight (lb)*	*Warhead (lb/kt)*	*Speed (Mach)*	*Range (nm)*	*Guidance*	*Launch platform and remarks*
1960 1962	SS-N-2A 'Styx' SS-N-2B 'Styx'	21	5,100	1,100lb	0.9	16	Active radar homing	'Komar', 'Osa'. SS-N-2B is export version
1967	SS-N-2C 'Styx'	19	5,500	1,100lb	0.9	45	Mid-course update, active radar homing	'Mod Kashin', 'Kildin'
1962	SS-N-3 'Shaddock'	36	10,000	2,200/200	1.4	250	Mid-course correction, active radar homing	'Kynda', 'Kresta-I, 'Echo-I', 'Echo-II', 'Juliett'
1970	SS-N-7	23	6,400	2,425lb or nuclear	Subsonic	35	Active radar homing	'Charlie'. Submerged launch
1969	SS-N-9	30	7,300	2,425lb or nuclear	0.9	75	Active radar homing	'Nanuchka', 'Sarancha'
1973	SS-N-12 'Sandbox'	36	10,000	2,200/200	2.5	300	Mid-course correction, active radar homing	'Kiev', 'Slava'
1981	SS-N-19	?	?	?	Supersonic	300	Mid-course correction possibly by satellite, active radar homing	'Kirov', 'Oscar'. Terminal manoeuvres incorporated
1982	SS-N-22	?	?	?	Supersonic	70	Mid-course correction, possibly by helicopter	'Sovremennyy'

Note: A proportion of any of the above types of missile may be fitted with infra-red homing heads, and some of the later models may home on radar transmissions.

exploitation is a less planned and clear-cut mission.

Allied to this modification of land warfare comes a dictum to which no close land parallel can be found. This is Gorshkov's often-quoted reference to 'the battle for the first salvo'. It can be inferred from this that the Soviet maritime forces expect, at some stage in the transition from tension to war, an order to launch a general, co-ordinated attack on designated maritime targets, and that their prior manoeuvres will be designed to get them into the best situation to carry out this attack. Had Gorshkov's been a chance remark, that inference might have been unjustified; but there is a great deal of evidence from other Soviet writings and, perhaps more importantly, from Soviet exercises, that the co-ordinated 'first salvo' is an important part of Soviet maritime doctrine, even though exercises in the mid-1980s have indicated a less rigid adherence to it, with the admission that a more graduated and flexible approach to escalation may in the event be necessary.

The necessary co-ordination for the 'first salvo' is not the easiest thing in the world to achieve. Even in the relatively restricted sphere of a regimental attack by missile-armed 'Badger' aircraft it demands accurate information on the designated target, derived probably from SOSS; localisation and tracking from the regiment's own airborne reconnaissance, probably 'Badger-D', 'E', 'F' or 'J' and probably listening for the opponent's transmissions rather than making their own; splitting the attack aircraft into several sections so that they launch their missiles from different directions, timed to arrive at the same moment; and doing all this with the minimum of chatter on the air waves, to avoid warning the enemy.

When to that relatively simple set-up is added the co-ordination of attacks from submarine and surface units, with the added complications of the relatively long time taken to manoeuvre them into suitable firing positions and communication problems (particularly with submarines), the question of target allocation against opposing units which may well be changing their formation or disposition relative to each other, and the rather limited approach lines imposed by the geography of both the Atlantic and Pacific theatres, the 'battle for the first salvo' is indeed a battle not only against the alertness of the opponent but against formidable problems of organisation.

But there is little doubt, in this writer's mind at any rate, that they are determined to do it, at least in the situation of an advance by major Western naval forces into the Norwegian Sea or Sea of Japan in a period of high tension. There has been criticism of NATO, much of it justified, for concentrating too much on a 'single scenario' concerning a Soviet attack on the Central Front; we sometimes forget that the Soviets also often appear to be obsessed with a single scenario, though in their case it is an approach to their 'bastions', as outlined above. How they would react to a less classical situation is by no means so certain; how well, for example, would the principles outlined above serve in the South China Sea or the Indian Ocean? It seems likely that the Soviet Navy would try at least still to apply them; how well it would succeed is open to doubt.

For, in effect, the co-ordination problem is an information-gathering and exchange problem; and in the Soviet case it depends on the efficient operation of SOSS and very tight control of all maritime forces from a strong and superbly-informed central headquarters. That organisation is undoubtedly optimised to deal with a threat to the vitals of Soviet power, as they are seen by the Kremlin. It may, in its present state, be barely sufficient to deal with that situation, but the best technology available to the Soviet Union is no doubt being pressed into service to improve it. For dealing with more outlandish threats, which may not be held to affect the Soviet Union so vitally, the organisation would almost certainly manage a far less precise measure of control, and much improvisation and initiative from commanders at sea might be needed. This has not in the past been a very strong point with the Soviet armed forces. However, the provision of command facilities at sea, and some training in their operational use, has been a notable feature of the Soviet Navy in recent years and it may be that the 1990s will see more independence and flexibility for sea-going commanders.

For the present, not much is known about the Soviet potential for exploitation of any initial success, or regrouping and re-entering the conflict after initial failure (but for that matter, how many Western exercises address the problem?). Nevertheless, some speculation about Soviet actions in such circumstances is necessary, if only to answer two important questions: how soon might they employ nuclear weapons, and how soon and how heavily might they conduct air attacks against reinforcement and resupply shipping crossing the Atlantic?

With regard to nuclear weapons, the best short answer from Soviet writings appears to be 'as soon as they are necessary to gain a decisive advantage'. This is not quite so platitudinous as it sounds. Twenty years ago the answer might more nearly have been 'at once'. Soviet thinking has become more flexible since then and it is possible that the 'first salvo', if achieved, might be conventional only, particularly if it was directed at forces whose threat to vital Soviet fleet or state functions was neither immediate nor overwhelmingly heavy. But of their intention to make tactical use of nuclear weapons at sea when they perceive a military necessity to do so, there can be no doubt.

So far as attacks on shipping are concerned, it is necessary first to distinguish between shipping in the Norwegian Sea and Sea of Japan — which must by definition be an important target — and transatlantic shipping. It appears from the low emphasis given to it in Soviet literature that the latter is not a high priority for the Soviet Naval Air Force. Missile-armed surface and submarine units are more likely to find such shipping among their targets but, even then, probably not on a large scale at the earliest stage of a conflict. If, later on, more general targetting of transatlantic shipping was ordered, the corollary given present Soviet doctrine would be that the principal threat nearer home had been largely eliminated. But one must enter a powerful caveat here. Doctrine, as has been apparent over the past decades, can change, and it is possible that the Soviet Union could re-assess the priorities so that Allied reinforcement and resupply of the Central Front became a prime target. If they made this re-assessment, its implementation is mostly a matter of deployment; the means are already in existence.

The Non-Soviet Threat

Over 70 nations can now pose an air threat to seaborne units. Mostly the threat is posed by a combination of shore-based aircraft and missile-armed fast attack craft; in the majority of cases, therefore, it is of relatively short reach — say only a few hundred miles from the bases of the respective threat units. Mostly, again, the missiles involved are of short range compared with those mounted by the Soviet Union or, more lately, by the USA. Overwhelmingly, the threat is non-nuclear.

Given these generalisations, the diversity of systems is very large. The Western technical press is full of advertisements for ship-attack systems, all claiming singular properties of accuracy and lethality; and as is well known, without similar *réclame* the Soviet Union has supplied anti-ship missile systems to a couple of score nations around the world. An attempt has been made in the table on page 33 to express this proliferation; but, to sharpen the perception, it is worthwhile to conduct a mini-case study of two nations that are actually in the ship-attack business as this book is written: Iran and Iraq.

The Gulf War began in September 1980 and most of its activity has been on land. But a quite important element of it has revolved around oil

Above:

Harpoon, here seen fired from the USS *Ticonderoga*, has been fitted increasingly in third-world navies, including some at war with each other in the mid-1980s.

US Navy Official by Bruce Trombecky

traffic in the Gulf, proceeding both to and from Iranian ports (particularly Bandar Khomeini and Kharg Island) and the northwestern Gulf (particularly the Iraqi installations off Fao). It is fair to say that because of pipeline arrangements, Iran is somewhat more vulnerable to interruptions in oil traffic than Iraq.

During the six years of war there have been many attacks by missile on merchant vessels, mostly tankers, engaged in this traffic. Before 1986 they were confined to the northwestern Gulf; however, in that year they were extended almost as far as the Strait of Hormuz. Most of the attacks have apparently been conducted by aircraft. The number of merchant ships actually sunk, relative to the number of attacks, has been remarkably small, but damage has been considerable and casualties significant. Insurance rates for the traffic are now extremely high. There have also been several fierce engagements between naval forces resulting in losses on both sides of surface craft, reportedly running into double figures.

It was not until 1987 that the first naval casualty occurred. USS *Stark*, a frigate on patrol in the Gulf, was attacked by a missile from an Iraqi aircraft and was severely damaged with the loss of 37 lives. In subsequent weeks Western naval pressure was strengthened and merchant ships closely escorted.

The forces conducting this war are not untypical of those of medium-sized powers with local reach, and (partly because of historical factors) both sides have drawn them from a variety of sources.

Iran, having been a client of the USA for many years while the Shah ruled, still has a Western slant to its force structure. The major surface ships, some seven in number, are all British or US in origin, and the dozen or so fast attack craft are French. They deploy a variety of surface-to-surface missiles. The destroyers are armed with Standard SM-1, a US system fitted primarily for air defence but having a secondary surface-to-surface mode. The Standard homing system is semi-active, that is to say the parent ship illuminates the target with its own radar and the missile homes on the reflected pulse. This limits the range to the ship's radar horizon, say 16 miles or so. However, there is also an anti-ship version of the Standard missile where the missile homes on the target's radar emissions; this would increase the theoretical range. It is not known whether the Iranians have made any use of this. The frigates are armed with a first-generation Italian missile, the Seakiller. This homes by riding down a beam transmitted by the parent ship. It, therefore, is also limited to horizon range. The fast attack craft have the US Harpoon. This has the attributes of sea-skimming, inertial mid-course navigation, autonomous active radar homing, over-the-horizon range (about 60 miles) and terminal manoeuvres designed to maximise confusion and damage to the target: in fact, a thoroughly modern missile.

Right and below:
Exocet is a sea-skimming missile now widely fitted in Western and third-world navies. It can already be ship-launched and air-launched, and a submarine-launched version is coming into service. *Aerospatiale*

The Iranian Naval Air Arm is orientated towards anti-submarine warfare, but the Iranian Air Force has some units — both fixed and rotary wing — which can deploy air-to-surface missiles. There appear to be two types in the inventory: the US Maverick missile whose terminal system is infra-red, homing on the emitted heat from surface craft, and the wire-guided French AS-12 missile normally fired from helicopters. The latter is of very short range, a mile or two at most.

It is also known that the Iranians have mounted on shore a Chinese-made missile, the Silkworm, capable of attacking ships at sea.

The Iraqi forces up to a few years ago were almost exclusively Soviet in origin, but since 1981 their procurement policy has become more catholic and they are acquiring a major surface force, aircraft and missiles from Italy and France, many of which will come into service as this book is written. Surface forces consist at present of 'Osa' class fast attack craft with the export model of the 'Styx', to be joined by a 10-ship frigate and corvette force armed with the Italian-French joint venture Otomat-II. This has very similar attributes to Harpoon, though its theoretical maximum range is somewhat greater and its speed, though still subsonic slightly higher.

There is no Iraqi Naval Air Force as such, but the Air Force has a considerable armoury of anti-ship delivery systems. There is a squadron each of 'Badger' and 'Blinder' aircraft said to carry either 'Kitchen' or 'Kelt' missiles, though only in the conventional-warhead version; and newly into service is the air-launched version of the French Exocet, a missile which needs little introduction after the Falklands campaign. It can now claim to be the most battle-proven of all ship-attack missiles (though the makers of the British Sea Skua, which scored eight hits out of a possible eight in the Falklands, might dispute that point). The Exocet is a sea-skimmer of about 40 miles range with active radar terminal homing and a 165kg (364lb) warhead. It was one of these that severely damaged USS *Stark*.

It must be pointed out that from the accounts we have of Iraqi attacks on merchant ships, the missiles used have not all been of the types described above; some have been a good deal smaller and less destructive. They may have been aircraft rockets, or air-to-air missiles used in an air-to-surface mode.

One form of anti-ship weapon system possessed by neither Iran nor Iraq is the submarine-launched missile. The waters of the Gulf are regarded as unsuitable for submarine operations and neither side has them. In the context of this book and at the present time, that makes the Gulf states less untypical than might at first appear. Of the 30 or so non-NATO and non-Warsaw Pact states that at present possess submarines, so far only Japan and Australia have firm plans to fit subsurface-to-surface missiles. Both these will be of the Harpoon type (the versatile Harpoon has an air-launched version as well). The missiles fit into a standard torpedo tube and the submerged-launch system is well proven. Limited though the distribution of such systems is now, it is doubtful if that will continue. It is hard to imagine Argentina, for example, remaining long content without its modern force of small German-built submarines being armed with such missiles. Whether those navies that have Russian or Chinese-built submarines will have the same aspirations — and whether suitable missiles will be made available to them — is more conjectural, but it would be surprising if there was not a tendency in that direction.

Above:
The helicopter-based Sea Skua, a weapon that proved to have an excellent reliability rate in the Falklands campaign. *British Aerospace Dynamics Group*

Below:
The main air threat as seen by the Soviets: A-6E Intruder attack aircraft sit with an EA-6B Prowler and S-3A Viking as an F-14 is launched from the USS *John F. Kennedy*. *US Navy Official*

Through Soviet Eyes

Finally, it is necessary to attempt to view the air threat to Soviet surface units as they may see it. In material terms it is of a different kind from that which they pose to the West. They will see the possibility of concentrated attacks by relatively small, very fast, agile carrier aircraft armed with missiles of no more than medium range but considerable sophistication and diversity in their homing systems. Terminal manoeuvres designed to confound defences are a feature. They will be impressed by the at-sea command and control systems available to American CVBGs and the flexibility these allow.

They may have been comforted, up to a very few years ago, by the fact that the Americans (much less any other NATO nation) deployed neither

very long range nor submarine-launched missiles. Now this comfort is denied to them on both counts. The Tomahawk cruise missile has a ship-attack version with a range of about 300 miles and will be very widely deployed in surface ships down to the 'Spruance' class destroyers and in submarines. In the latter it will be fired from specially-fitted vertical launch tubes. The missiles are subsonic but incorporate special technology to make them difficult to detect by radar. They have active radar terminal homing. The Tomahawk does suffer from the well-known limitations connected with target acquisition and mid-course guidance, but the Soviets will be impressed with the US emphasis on information technology and command and control and may well judge that the cruise missile is a very important addition to the US armoury.

Moreover, that is not the Soviets' only recent headache. The advent of the Harpoon in air, surface and submarine-launched modes fills the middle ground between the very long and very short ranges all too well. Fitting in the US maritime forces is now very general, and the threat, particularly if attacks are co-ordinated, severe.

As if this were not enough, the other NATO navies face the Soviet Union with an unpleasant diversity of systems, for they have by no means all standardised on the US Harpoon. The widely-fitted French Exocet — available in air, ship and submarine-launched versions — the Otomat, the Norwegian Penguin, the British air-launched Sea

Below:
A diversity of anti-ship missiles faces the Soviet Navy. One such is the Norwegian Penguin, seen here in Mk II and Mk III versions. *Kongsberg A/S*

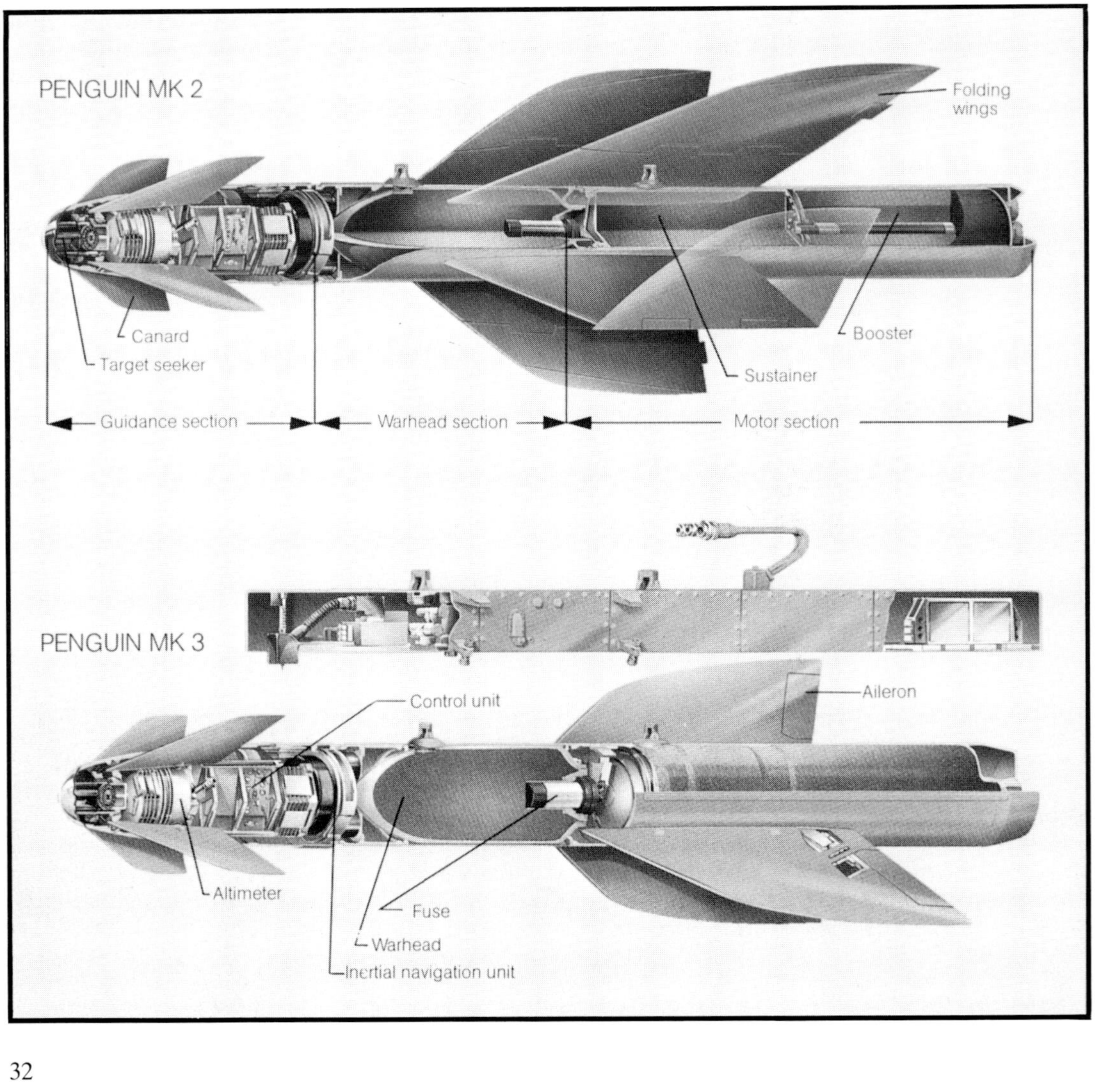

NON-WARSAW PACT ANTI-SHIP MISSILES

Country of origin	Designation	Length (ft)	Launch weight (lb)	Warhead (lb)	Speed (Mach)	Range (nm)	Guidance	Launch platform and remarks
France	Exocet MM38	17	1,600	364	0.93	26	Inertial, active radar homing	Surface ships of many navies. Sea skimmer
France	Exocet AM39	15.4	1,430	364	0.93	42	Inertial, active radar homing	Aircraft
France	Exocet SM39	19	?	364	0.93	32	Inertial, active radar homing	Submarine
France	Exocet MM40	19	1,880	364	0.93	45	Inertial, active radar homing	Surface ships
France/Italy	Otomat-II	14.6	1,700	440	0.9	100	Inertial, active radar homing	Surface ships. Sea skimmer with terminal bunt and dive
Israel	Gabriel-I	11	950	165	0.7	13	Beam riding	Fast attack craft
Israel	Gabriel-II	11	1,150	165	0.7	20	TV head, command guidance	Fast attack craft
Israel	Gabriel-III	12	1,230	165	0.7	25	Active radar homing	Fast attack craft
Italy	Seakiller-II	15.4	660	100	0.8	16	Beam riding	Iranian frigates
Norway	Penguin-1	10	730	?	0.7	14	Infra-red homing	Fast attack craft
Norway	Penguin-2	10	750	265	0.7	15	Infra-red homing	Fast attack craft and helicopters
Norway	Penguin-3	10.5	770	265	0.8	20	Infra-red homing	Aircraft
Sweden	RB-08A	19	2,700	?	0.85	120	Active radar homing	Coast defence
Sweden	RBS-15	14.3	580	?	0.9	60	Active radar homing	Aircraft, fast attack craft, submarines
UK	Sea Skua	8	165	45	0.8	9	Semi-active on helo radar	Helicopter
UK	Sea Eagle	13.5	?	?	0.9	50	Active radar homing	Fast attack craft and aircraft. Selectable trajectory
USA	Harpoon	15	1,475	500	0.9	60	Active radar homing	Aircraft, ships and submarines. Terminal manoeuvres
USA	Tomahawk	20.5	2,430	1,000	0.8	300	Active radar homing	Ship attack version. Widely fitted in US ships and submarines
USA	Shrike	10	390	145	2.0	15	Home-on-radar	Aircraft
USA	Maverick (Navy)	8	633	299	?	?	Infra-red	Aircraft

Eagle and Sea Skua, to say nothing of various surface-to-air systems in a secondary surface-to-surface mode, all have their singular characteristics and could give the Russians some unpleasant surprises.

Moreover, the ocean deployments of the Soviet Navy must lead them to assume some threat from Western-orientated navies such as Japan and from third-world navies. Clearly they will hope that the deterrent effect of their overall military power will serve to prevent any attacks, but they cannot ignore the possibility, and nothing in their past planning or practice suggests that they will.

Finally, the Soviets will be looking carefully, and probably with their customary suspicion, at the apparent Western emphasis on the conventional end of anti-ship warfare. Remarkably few Western systems are advertised as able to take a nuclear warhead, and the whole thrust of Western development has been towards the economical, the accurate and the clever. The inference that might be drawn is that at sea, the Western intention is to keep the conflict conventional as long as possible, with nuclear weapons held to be delivered as a last resort and then only by the crude means of the free-fall bomb. The Russians may not, however, draw that inference. More likely, they will just not believe the tactical nuclear capability is that limited; or they will reason that in another medium, that of underwater warfare, the West will more quickly be forced into nuclear means and that this will unleash general use of nuclear weapons at sea. The evidence continues to suggest that, although they have built considerable conventional flexibility into their ship-attack systems, their basic assumption for major conflict at sea is that it will eventually involve the use of nuclear weapons.

CHAPTER 2

The Means

The objective of air defence at sea is to preserve, in the face of airborne attack, the effectiveness of surface units so that they can carry out their assigned missions.

Much has been written, particularly in recent years, to suggest that this aim is now unattainable, or is attainable in such limited circumstances that it is unrealistic and uneconomic to pursue it. Such questions require, for their resolution, strategic net assessments of enormous complexity. It is doubtful if an analysis based on all the relevant facts could come to a firm conclusion; and no single body has access to all the relevant facts anyway; and this writer is certainly not going to attempt anything of the sort.

It does need to be said, though, that many of the operations of maritime power require absolutely the use of surface units for their execution. These include a very high proportion of constabulary tasks in normal conditions; almost all low-intensity operations involving demonstration of right or resolve, and control work against piracy, gun-running, terrorism or illegal incursion; all major overseas reinforcement or intervention operations; all operations now categorised as power projection, when that power is air-launched or amphibious; and all protection of seaborne trade or resupply. It is not unreasonable to require, from those who say that surface units can no longer be sustained in the face of an air threat, a statement as to how military operations may be brought to a successful conclusion in their absence. So far, no statement that satisfies this requirement has been forthcoming.

The underlying strategic assumption of this book, therefore, is that the objective of air defence at sea is a necessary one, and will be necessary in a wide variety of circumstances.

The Impact of Technology

'Technology Rules' is a slogan so obviously to be applied to the air defence problem that it needs to be examined with care and, even, some scepticism. Certainly the first two chapters of this book have indicated how heavily technology impacts upon every aspect of the business; this and subsequent chapters will bear that out. But technology must be compatible with the physical laws of the above-water environment, and it must be worked by people.

The Above-Water Environment

The air above the sea is a much simpler environment than the sea itself. In a previous volume in this series on *Anti-Submarine Warfare*, discussion of the vagaries and uncertainties of the sea, and their effect on the sonic sensors that rule there, covered several pages. Concerning the air, it is tempting to say that there are only a few variables: it can be hot or cold, wet or dry, clear or cloudy, and none of these variables makes anything like the difference to the main sensors that water conditions do underneath. Even weather factors like heavy rain or rough seas become critical only in certain sets of circumstances. This is not to say that environmental variables do not produce *some* anomalies; they do, and due weight will be given to them later in this chapter. But they are not governing factors in the same way that water conditions are in ASW.

The factors that do govern the above-water scene are, much more, the laws of geometry and physics. First and foremost, nearly all methods of telling what is going on are to do with radio or light waves, which travel at immensely fast speed. Some means of processing the abundance of data is therefore needed. Moreover, these waves, with the exception of the lowest radio frequencies, travel in straight lines. This means that horizons are of great importance; the curvature of the earth is not negotiable. Finally, the behaviour of electromagnetic waves when confronted with reflecting or absorbing surfaces is of critical importance.

The laws of physics also govern the vehicles despatched, both towards surface units with the intention of doing them harm, and away from them with the intention of nullifying or mitigating

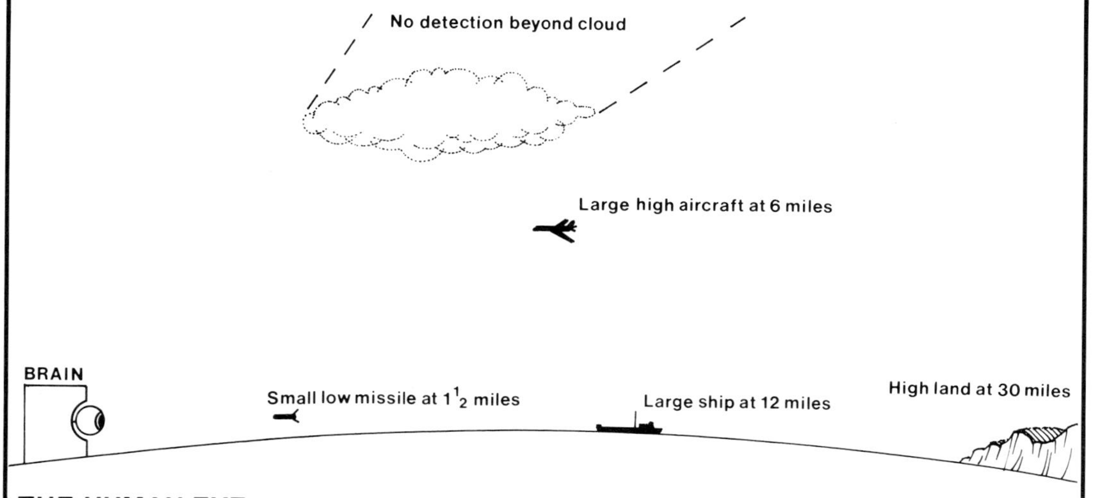

**THE HUMAN EYE:
TYPICAL DETECTION RANGES IN CLEAR CALM WEATHER**

the harm. Airborne vehicles have to overcome gravity, drag and the heat generated by friction. They must be able to manoeuvre with sufficient agility to accomplish their tasks. It has already been suggested in Chapter 1 that in the case of sea-skimming missiles, for example, the step-jump from subsonic to supersonic speed is a considerable one; in the same way, very fast and manoeuvrable air-defence missiles do represent a quantum advance on those of an older, slower generation.

Sensors

The Human Eye

The human eye operates in a limited part of the light spectrum bounded by infra-red at one end and ultra-violet at the other. It cannot penetrate opaque objects such as clouds or the sea, and it needs a good deal of ambient light to operate at all. Its angular resolution is very good but it lacks precise vertical and horizontal planes of reference, so requires some form of machinery to translate its data into accurate measurements of elevation and bearing. Its intrinsic ability to measure range is almost nil. The eye's chief virtue is its direct connection to a processor with unique decision-making and intuitive properties, the human brain. This enables it, for example, to alert other sensors which have been looking in the wrong place; or to modify information that other sensors without these unique properties have provided; or to make up, to some extent, for the eye's own physical inadequacies by the provision of estimates based on memory and experience.

In any discussion of the means of air defence it is quite important to consider the eye in those analytical terms. A great deal of claptrap is talked in such terms as 'The best sensor is still the good old Mk 1 Eyeball'. The eye does have great virtues, as have been described, but it also has considerable limitations not only in its range and scope but also in circumstances where it is desired to interface it to the weapon systems that can alone bring off effective air defence.

The eye is also, of course, a principal gateway to human perception and this is a critical factor in the business of controlling air defence, particularly in complex situations. But discussion of that should come later in this chapter.

Radar

The basic principle of radar is a quite simple one. A very short pulse of energy at radio frequency is sent out from a transmitter via an aerial, strikes an object and is reflected by it. The reflected pulse is picked up by a receiver (usually in the same unit as the transmitter). The elapsed time between transmission and reception is measured electronically; knowing that radio waves travel at the speed of light makes it easy to translate this into range.

Refinements crowd in. Accurate location of a target in airspace requires not only range but bearing and elevation. Bearing is found in most radars by shaping the aerial array so that the pulse is transmitted in one direction only; before the next pulse goes out the aerial turns a fraction, and so on throughout a complete rotation of the aerial, assuming all-round cover is required. It is important to ensure that the received pulse is related, throughout its processing, to the corresponding transmitted one, otherwise ambiguities arise, and also to ensure that no target should be missed; even in the simplest radar sets this means careful balancing of pulse repetition frequency, aerial rotation rate and the angular width of the beam.

Elevation presents a more difficult problem. In primitive radars it was estimated by turning to account a phenomenon of radar that is of some general relevance. The pulse that strikes a target follows two paths, not one; the first is direct, the second is first reflected off the sea's surface. The returned pulse will be a combination of reflections from these two. If they are in phase with each other, the signal will be strong; if out of phase, it will be weak or even zero. Where these maxima and minima occur depends on the characteristics of the particular radar set, but it will be directly

Left:
The directional rotating aerial for a modern search radar, the Sea Tiger. *Thomson-CSF*

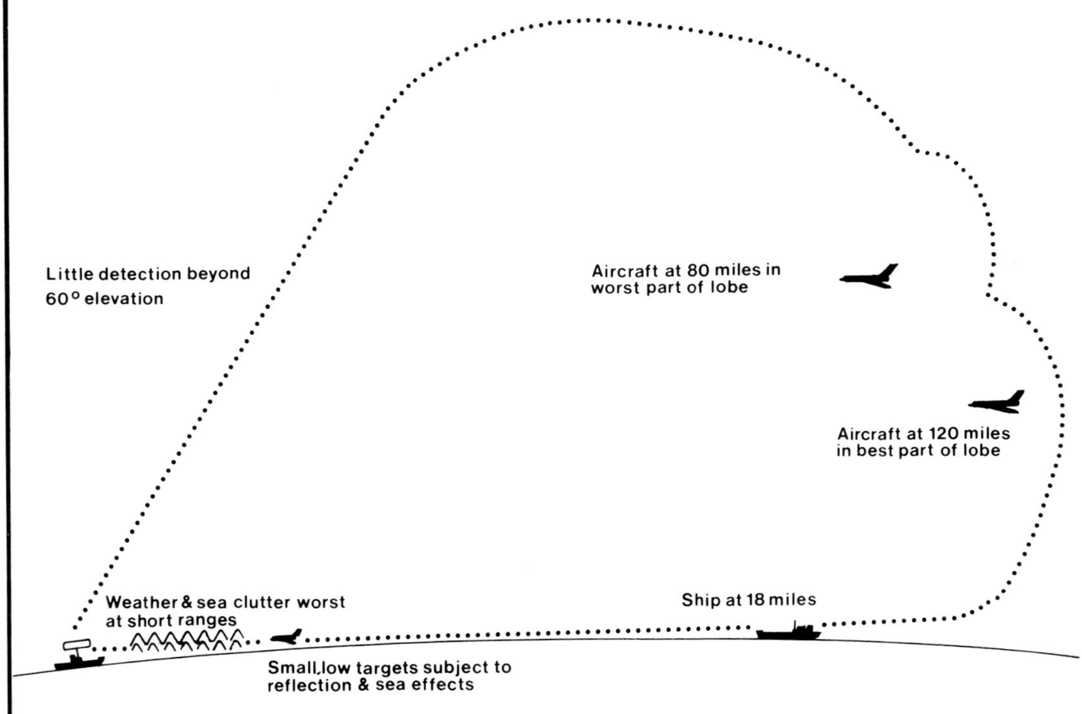

RADAR: TYPICAL CHARACTERISTICS & LIMITATIONS OF SMALL-SHIP GENERAL PURPOSE EQUIPMENT

related to the elevation of the target and, by the skilful use of the 'vertical coverage diagrams' that could be generated, fair estimates of elevation and therefore height could be made.

In modern conditions this was altogether too crude and slow, and other means of finding elevation were sought. It was, clearly, possible to shape the aerial array so that the beam was thin in the vertical as well as the horizontal plane; but this would greatly increase the time needed to search the whole sky. In some systems, a special radar of this sort was dedicated to height-finding. A more physically complex solution was to provide each aerial array with a whole battery of thin beam emissions set at different elevations; this meant a heavy aerial array, but did provide instant height-finding; such systems are commonly called three-dimensional or 3-D radars. Finally, the electronic agility conferred by modern printed circuitry and transmitting devices has allowed the introduction of electronic scanning, whereby a multiplicity of beams can be directed in both elevation and azimuth, ambiguities resolved and 'dwell times' adjusted to ensure adequate target tracking while the radar continues its electronic scan — all at a price, both in initial cost and topweight.

The paragraphs above have concentrated on ship-mounted radar in the air warning role. Most of the principles apply to search radars in aircraft and to target indication radars in ships, though in both cases weight and space considerations assume even more importance. They also apply, sometimes in modified form, to the pencil-beam radars that are employed in weapon control — both in the missiles that threaten the surface units and in the systems by which these units seek to protect themselves and their consorts.

The enemies of radar are threefold: attenuation, unwanted echoes or 'clutter', and the opponent's electronic countermeasures (ECM).

Attenuation, the dissipation of radio frequency energy by absorption in the atmosphere, is probably the least of the problems. Increases of transmitted power, though expensive, are achievable partly through improved aerial design and partly through improved power sources.

Clutter is a much more intractable problem and is linked with the search for greater power. The more the transmitted power, the greater the number of unwanted reflections there will be from wavetops, clouds, rain, even birds. The echoes from wanted targets — aircraft and missiles — must be distinguished from these. One important way of doing this is to use the doppler effect — the frequency shift that occurs in the reflection from a target that is moving rapidly towards or away. Another is to build coherence into the processing gear: only targets that show consistency in their

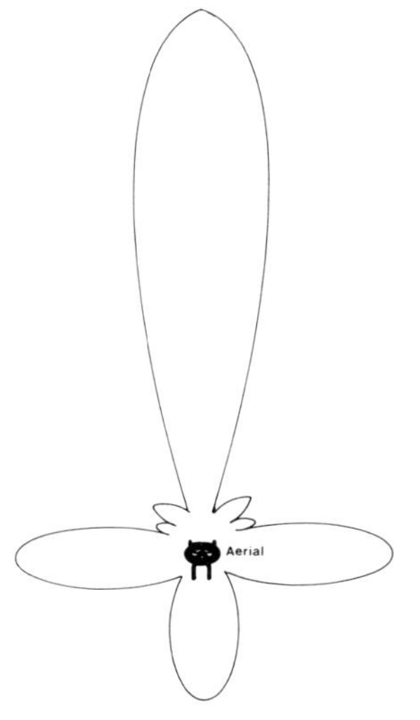

RADAR: HORIZONTAL COVERAGE OF ROTATING·AERIAL EQUIPMENT
ENERGY & UNWANTED ECHOES CREEP IN THROUGH SIDE & BACK LOBES, CONFUSING THE PICTURE PARTICULARLY UNDER JAMMING OR HIGH-CLUTTER CONDITIONS

movement throughout a succession of pulses are accepted. Finally, in the particularly difficult case of the sea-skimming missile target — where all the factors of a short horizon, wave clutter, reflections of the transmitted pulse from the sea resulting in multi-path reception, and a target of small radar cross-section conspire against the air defence radar — unusual combinations of frequency and aerial array may be tried.

ECM fall into two main categories: electronic jamming and physical means. Jamming can be over a broad or narrow frequency band, sophisticated or crude in its own frequency pattern. Its effect is to swamp the receiver with energy so that the wanted echoes cannot be distinguished. The chief counters in radar design are to improve aerials to eliminate the so-called sidelobes which enable jamming to creep through on multiple bearings; and to make the radar frequency-agile, so that the jammer can never fully lock on to a spot frequency and be at its most effective. Physical ECM is, typically, chaff (the direct descendant of the 'Window' of World War 2) dispensed either in clouds or more concentrated packages, the first to swamp the target radar with clutter, the second to seduce it into following false targets.

Lasers

Laser is an acronym for Light Amplification by Stimulated Emitting Radiation. The description is not immediately self-evident, and it must be said that if the basic principle of radar is simple, that of the laser is not. Certain substances, when

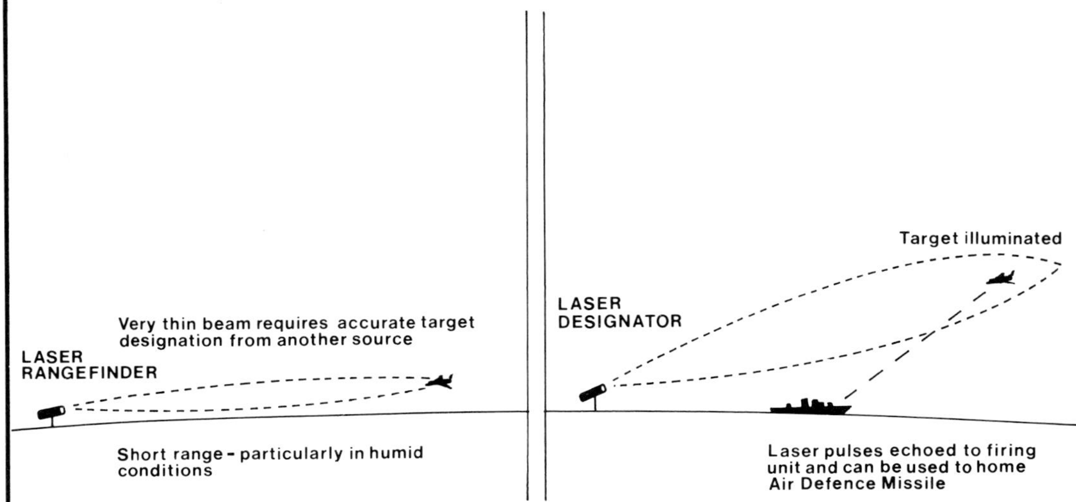

LASERS: Applications for rangefinding (left) and designation (right)

stimulated, excited or 'pumped' in certain ways, have the property of emitting very concentrated and powerful light spanning the narrowest of band widths. If suitable mirror or lens arrays can be devised, the emissions can be made highly directional, into beams, and if they can also be pulsed then the same effect as a pencil radar beam is achievable — only more so, so far as angular resolution is concerned.

Range, however, is another matter. Laser emissions are subject to attenuation from atmospheric conditions, particularly if these are humid, and to a phenomenon called thermal blooming, which is caused by the beam's changing the refractive properties of the air through which it passes, with consequent bending and diffusion.

Another drawback to the laser is its potential effect on the eyes, not of opponents but of friendly forces. Some sorts of laser beam, concentrated on the retina, can cause permanent blindness even if transmitted at very low densities. It is one of the oddities of lasers that other sorts of beam, not too distant in frequency, meet enormously higher tolerances; for example, a ruby laser is 200,000 times as dangerous as an erbium/glass laser in this respect. But whatever the laser, care is needed.

Lasers are still a long way down the learning curve, the principle having been around for only 30 years and the technology for a much shorter time than that. Yet they are already in widespread use as components of 'smart' systems, particularly as illuminators for semi-active homing weapons; and are on offer as weapon control systems for shipboard air defence. Their advantages in precision and anti-clutter properties may well in some circumstances outweigh their range limitations.

Lasers also have potential as weapons in their own right; this will be discussed later.

Infra-Red

With infra-red we move perceptibly towards those sensing systems which, rather than send out energy and detect the reflections, actually use emissions that the target itself makes. For while there are some infra-red systems that actually emit beams which are then reflected, either to the emitter or more usually to a sensor in a missile — the now-familiar semi-active mode — most of them, whether in platforms or the homing heads of missiles, simply detect the infra-red light that is already there, emitted by ships, submarines, aircraft or missiles.

The advantage of such a 'passive' system is that it is itself undetectable, and moreover it uses very little power compared with active or semi-active systems. But it does depend, not perhaps on the co-operation of the opponent — all targets are heat-producers and, however hard they try to minimise their consequent infra-red signatures, detectable to a degree — but on the circumstances of the time. In particular the efficiency of the system at any moment will depend on the target's aspect; as an example, it is clear that the heat and

Right:
A television/infra-red tracker. *Thomson-CSF*

therefore infra-red emission from the jet-pipe of an aircraft or missile is many times greater than anything presented by its head-on aspect.

In consequence, in the design of any infra-red equipment a careful balance has to be struck between insensitivity — which will miss out targets below a certain threshold of emission, and that might include head-on targets that one dearly wanted to detect — and over-sensitivity, with its attendant risks of high false alarm rates and susceptibility to seduction either by deliberate action of the opponent or through atmospheric anomalies.

Electronic Support Measures (ESM)

Of all the gabble-phrases used in maritime warfare, ESM is one of the most irritating. It does not say what it means. The excuse of those that coined it is, presumably, that what it means is quite complicated. In brief, it refers to the detection and exploitation, for purpose of information and warning, of the opponent's emissions in the electromagnetic spectrum.

It has been shown above that of the several forms of positive action that can be taken to discover the position and movement of opposing maritime units, most require the transmission of packets of energy, often of high power, in the expectation that it will be reflected and provide information for processing and tactical use. The most generally useful system is active radar; it is of longer range, more reliable and more easily processed.

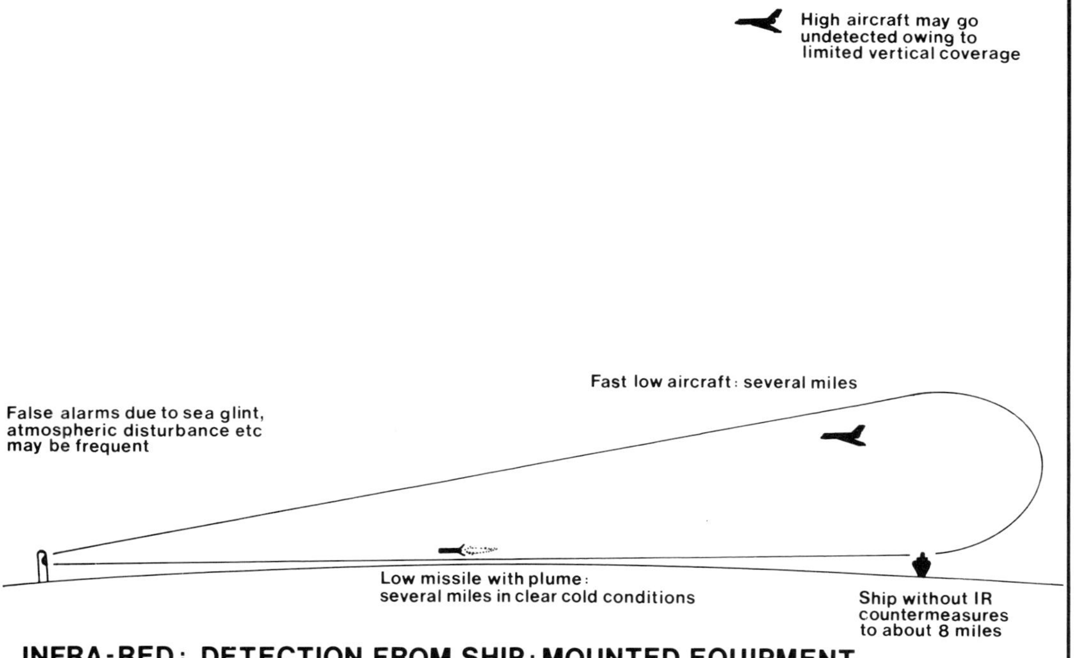

INFRA-RED: DETECTION FROM SHIP-MOUNTED EQUIPMENT
IR is often, per contra, used in missile heads for target detection or terminal homing

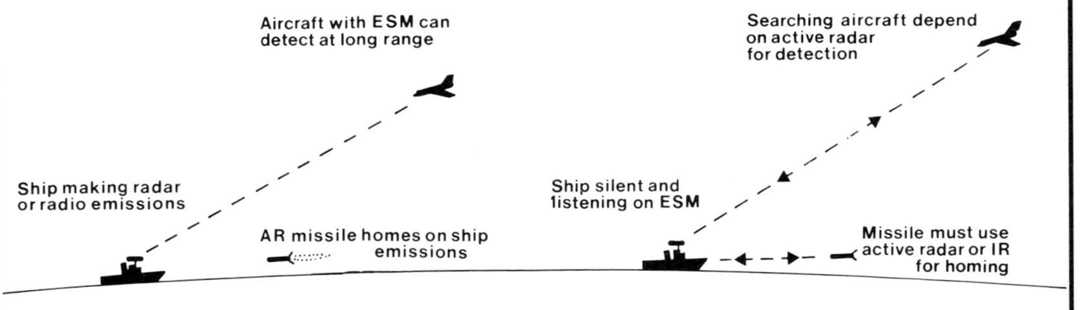

EFFECT OF ELECTRONIC SUPPORT MEASURES

Like any other radio emission, active radar emission can be detected by a receiver. The receiver must of course be capable of accepting the transmitted frequency: the narrower the band of acceptance, the more sensitive the receiver will be, but the greater will be the possibility that the radar transmissions will fall outside the frequency scope of the search. By making the receiver directional, which is a fairly simple matter in these days of advanced aerial design, the bearing of the received emission can be measured.

There is no way of measuring range from a single receiving point. But if simultaneous, or nearly simultaneous, detections can be made from two or more units, then the position of the emitter may be determined by the use of cross-bearings. Good communications between the detecting units will be needed. (That will not be the last time that this particular requirement appears in this book.)

ESM techniques can also be used in homing systems themselves. It is easy to devise a missile head that homes on an opposing radar, and the more directed the radar's beam and the more powerful it is, the more the missile likes it. The head must of course be tuned to roughly the right frequency, and this can cause design problems because of the small space available for sophisticated equipment.

There are two massive advantages to ESM as a means of detecting the opposition. First, it does not itself make an electronic emission and therefore does not give away one's own position. Second, it confers range advantage. Active radars, working on reflected energy as they do, have to use very high power to compensate for the loss of energy in the double journey and in the process of reflection itself. In consequence, the range at which their transmissions will be detected is almost invariably greater than the range at which they will detect the platform employing only ESM.

There is one massive disadvantage to ESM: it depends on the cooperation of the enemy. An opponent keeping strict radio silence will not be detected. How much harm that opponent can do while he keeps radio silence will, of course, depend on the tactical situation.

Strictly speaking, ESM covers the detection and exploitation of communication as well as radar systems. It is still possible to carry out such operations, notably against units of those naval powers that do not use satellites for communication; in particular, exploitation of the high-frequency band transmissions of any opponent unwise enough to use it prematurely can be highly productive.

The Active-Passive Balance

So the wheel comes full circle. A popular writer on bridge once said that in every hand there was a time to lie doggo and a time to get busy. So it may be in most air defence situations at sea. To rely overmuch on active systems, however effective they may be, risks the stealthy (and probably very economical) riposte. To rely too much on passive systems may risk being out-guessed and out-flanked by an even wilier enemy. A strong hand will contain a sufficient variety of both.

Weapon Systems

It is not necessary to discuss weapons systems individually here — that is for the next chapter — but as categories, so that a picture of what is available can emerge and some principles for air defence at sea can be derived. It is preferable to discuss platform/weapon combinations as single systems, since in practice they do tend to be integrated; nowadays only the simpler air defence systems can be 'strapped on', although that kind of characteristic is highly desirable and the prospects for its realisation ought always to be kept in mind.

Airborne Systems

Airborne air defence weapons are generally carried in fixed-wing combat aircraft, working either from shore bases or aircraft carriers. The aircraft have sortie times typically between one and three hours from take-off to landing; it is fair to assume that a sortie longer than three hours will involve in-flight refuelling. The missiles they carry are currently optimised to shoot down opposing fixed-wing aircraft.

Long-range (over 40 miles) missiles demand great capacity and sophistication in the parent aircraft as well as the missile. Operating usually at very high relative speeds of closing, the parent aircraft must obtain a firing solution in order to make best use of the missile's range; this can entail detection at over 100 miles. Then the missile must be placed in a position where it can effectively home on the target. But, of course, the exposure of the parent aircraft to counter-attack is minimised by the long range of the missile.

Medium-range (say 20 miles) missiles cannot themselves acquire their target before leaving the parent aircraft. They therefore need semi-active guidance throughout their flight to the target, or must be inertially guided until their own homing systems take over. Semi-active guidance means that the parent aircraft must stay in the vicinity of the target — not something aircraft tend to wish to do — while inertial guidance can increase the possibility of missing, particularly in the case of a crossing target.

Short-range (say three miles) missiles rely on acquisition when still attached to the parent aircraft. This gives a positive 'go' for firing to the pilot; but it entails good manoeuvring on his part, particularly as the preferred homing system is often infra-red and thus acquisition range depends on target aspect. And, of course, exposure of the firing aircraft is at a maximum.

So far, the ability of air-to-air missiles to shoot down other missiles is limited. The closer a target missile is to being a kind of aircraft, the more likely it is to be shot down; so the very large missiles favoured by the Russians may be somewhat more vulnerable to air-to-air attack than the relatively slimmer Western ones. On the other hand, they tend to be faster; and few things are more frustrating than a stern chase by a Mach 2 missile of one that is travelling at Mach 2.5.

Perhaps the most difficult target of all, for airborne systems in the current state of technology, is the sea-skimming missile. Although its

Below:
Air-to-air missile under the wing of an F-14A Tomcat aircraft on the flightdeck of USS *Carl Vinson*. *US Navy Official by PHAN Charles R. Solseth*

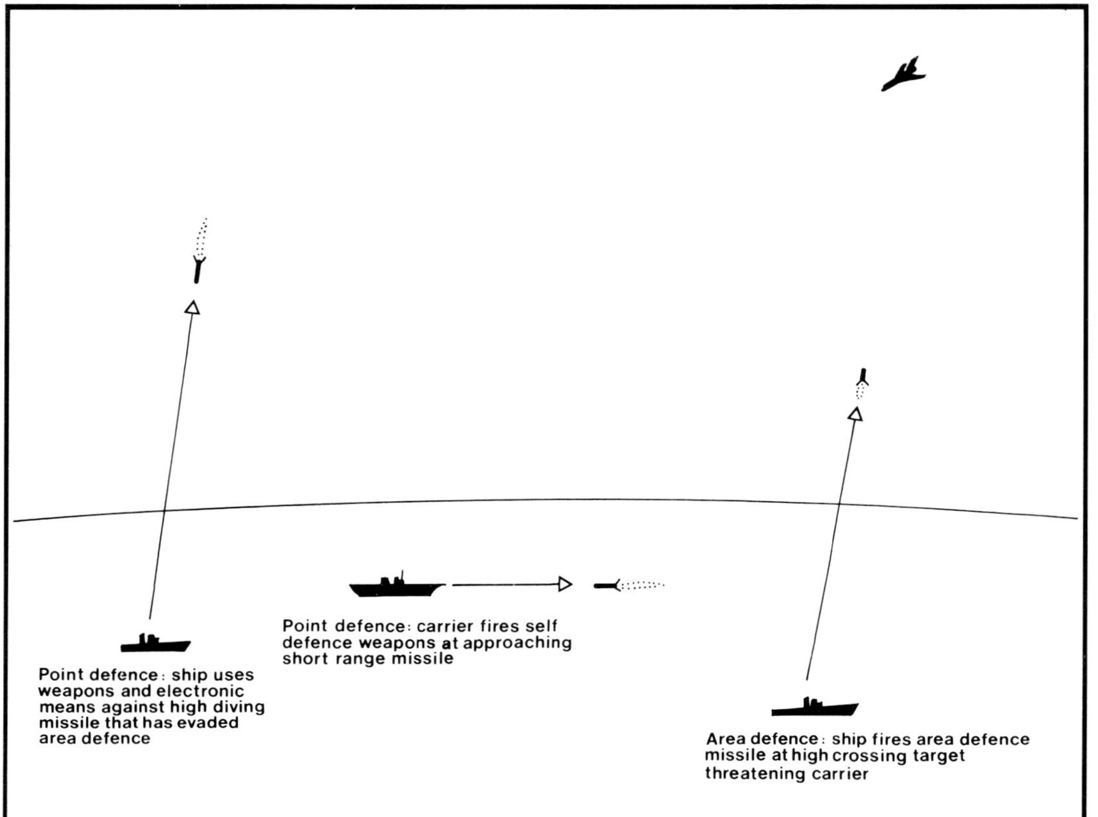

AREA & POINT DEFENCE: Typical applications

radar cross-section may be at maximum seen from above, that is about the only advantage it gives to the opposition. The great majority of detections will have to be unalerted, since no other unit (whether airborne or on the surface) will be within the limited detection range, and the time available for effective action will be short. Great efforts are being made, by both superpowers, to enhance 'look-down-shoot-down capability'; but it will always be a far cry from the deliberate, vectored approach to an enemy aircraft detected at long distance by ship or airborne early warning radars.

Before leaving the subject of airborne systems, it must be pointed out that the aircraft is a prime instrument in the destruction or deterrence of surface and sub-surface platforms, many of which would, if left undisturbed, pose an air defence threat. To this extent its contribution to the whole air defence scene is enhanced.

Shipborne Systems

Shipborne air defence systems are numerous and the category lends itself to sub-division. Generally they fall into hard-kill systems, which seek physically to destroy the incoming threat, and soft-kill systems which seek to divert it or make it commit suicide. The hard-kill systems can be further sub-divided into area defence, where a whole formation of ships may be protected by one source, and point or self defence, where only the system's own ship is designed to be protected.

Whatever system is being employed, the air defence sequence falls into phases: detection, tracking, acquisition and engagement. This sequence holds good whether the target is an aircraft or missile, and to a large degree whether the method used is hard-kill or soft-kill or a combination of both. The classification of the target as hostile is an important part of the process but can occur at varying times in the sequence.

Area defence is probably the most difficult problem, and certainly that which is most expensive to solve. To cover an extensive force of ships entails engaging targets which are changing their position in three planes, often at long range, at all heights from very near sea level to tens of thousands of feet and at all elevations from horizon to zenith.

Detection, then, must take place as far from the protected force as possible for two reasons: first, to ensure that outlying units of the force can come within the cover of the area defence system; second, to ensure as long a period of tracking as possible. The target then needs to be transferred to the fire-control radar which can track it more precisely; this process is called acquisition and may in some cases require a radar intermediate between the long-range warning and the fire-control sets. That will certainly be so, for example, if the initial detection and tracking are made by an airborne radar.

Once the fire-control radar is firmly tracking the target — in nearly all modern systems this is done automatically, leading to the familiar expression 'locked on' — the missile or missiles may be launched, as soon as their maximum range equates to the predicted range of the target at the point of meeting. The computation of this point is relatively easy to achieve electronically — so long as the target obliges by continuing on its previous course and speed during the flight of the missile.

Naturally, this cannot be relied on, and the missile must be made to home on to its target. In area defence systems in Western fleets, the preferred method of doing this is the semi-active mode, the parent ship's fire control radar illuminating the target and the missile's head sensing the reflection and homing on to it. The more illuminating radars that can be provided, the more targets can be engaged simultaneously. Information on Soviet area defence systems is not easy to come by, but it is reported that the more modern may employ terminal homing by active radar in the missile's own head, after mid-course guidance achieved through a missile-to-target-to-missile-to-ship link.

Several factors in the discussion above — in particular the need for early detection, the desirability of ample tracking, and the need for complex and clutter-prone guidance and homing systems — suggest that the ideal target for any air defence missile system is of fairly large size and

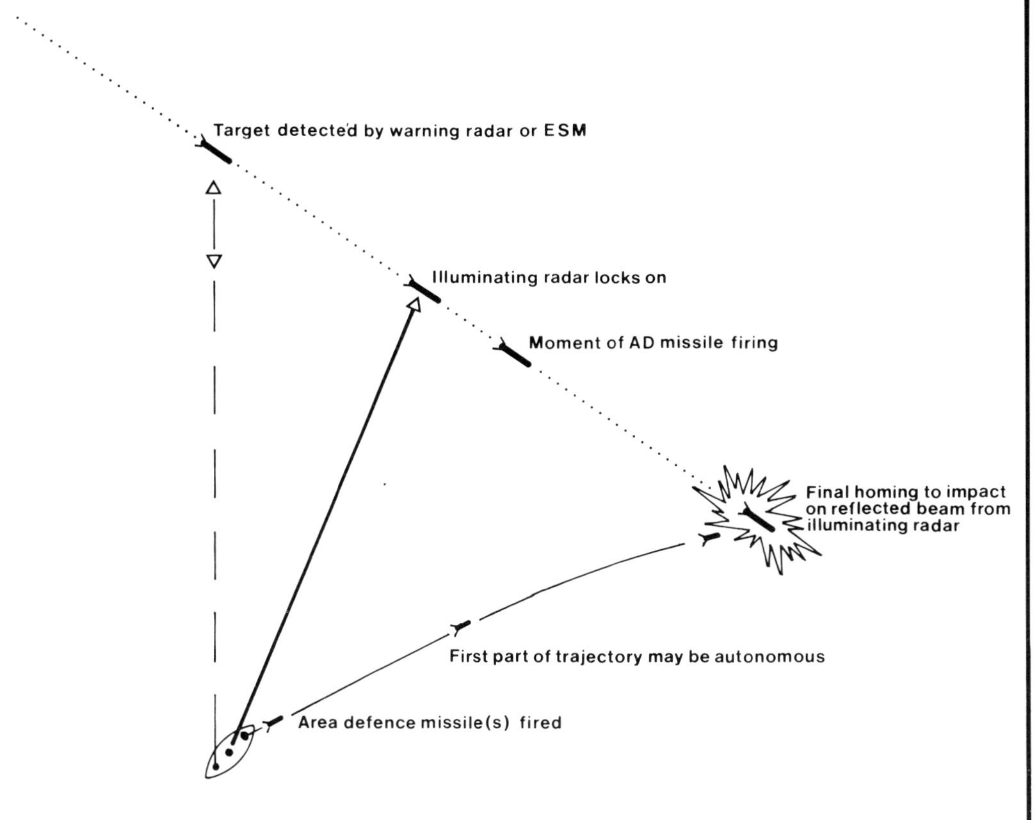

AREA DEFENCE: Engaging a crossing target, either an aircraft or large missile

flying at a moderate height. Once more, the sea-skimming missile, particularly if it is of small radar cross-section, emerges as a difficult target. So does the very high-flying, steep-diving missile.

Point defence systems are more numerous in kind than area defence systems, and employ a greater variety of ways of destroying the target — or hoping to do so.

Point defence missiles have an effective range — that is, the maximum range at which they can be expected to intercept a target — typically of well under 10 miles. This does not mean that the target's range when they are fired will necessarily be a whites-of-the-eyes job; if, for example, it is desired to intercept an approaching Mach 3 missile with a Mach 2 missile at its six-mile maximum effective range, then a fire control solution must be achieved and the point defence missile fired when the target missile is 15 miles away.

That kind of requirement could be virtually incapable of fulfilment against a sea-skimmer; but fortunately, as we have seen, most current sea-skimmers are not supersonic, and most missiles capable of Mach 3 are quite large and higher-flying, and therefore fire control solutions on them may be available at reasonable ranges. Nevertheless, it is as well to keep the example in mind, for it is a possible 'worst case' for the future and even under current conditions indicates that, given the inevitable lags in reaction time however smart the drill, most point defence engagements are likely to take place uncomfortably close to the ship.

For missile point defence systems a popular means of achieving interception is to lock a narrow radar beam on the target and then either command the point defence missile on to this line of sight or make it 'ride' up it electronically. The former method is the more flexible, since command-to-line-of-sight can be applied to optical or infra-red tracking systems as well. But for any such system, however the final guidance is to be achieved, initial alerting and acquisition are critical and time can be desperately short. In consequence, many modern point defence systems are designed to minimise the need for human reaction; the whole process, from warning through acquisition to lock-on and engagement, can be done automatically. This puts a high premium on reducing false alarm rates; too many of these, and ammunition will be uselessly expended at a great rate.

Missiles are not the only means of achieving point defence. Guns have made something of a come-back in recent years, in particular modern versions of the smaller-calibre guns that in the later years of World War 2 filled the sky with lead and gave much confidence to hard-pressed ships under air attack. The salient characteristics of air defence guns now are accurate pointing, a very high rate of fire and ammunition designed for maximum lethality. As always, the acquisition process is a critical one, but many modern systems have alternative means of fire control — radar, optical, infra-red — and the guns' rate of slew is very rapid, so the chances of engagement are fair. The very short open-fire ranges, typically a couple of miles, mean that an incoming target must be destroyed quickly; otherwise, even if mortally wounded, it may fly through and still hit the ship.

Finally, among the hard-kill weapons, it is perhaps right to mention the potential of the high energy laser. As early as 1978 a US system of this type destroyed, under test conditions, an anti-tank missile flying at subsonic speed. It can be assumed that in both superpowers great efforts are being made to transfer that technology into an effective shipborne system.

The advantages would clearly be tremendous. First and foremost, the ammunition would be the ship's main power supplies: not inexhaustible, but a great deal less limited than a magazine. Rapid re-engagement, without any reloading require-

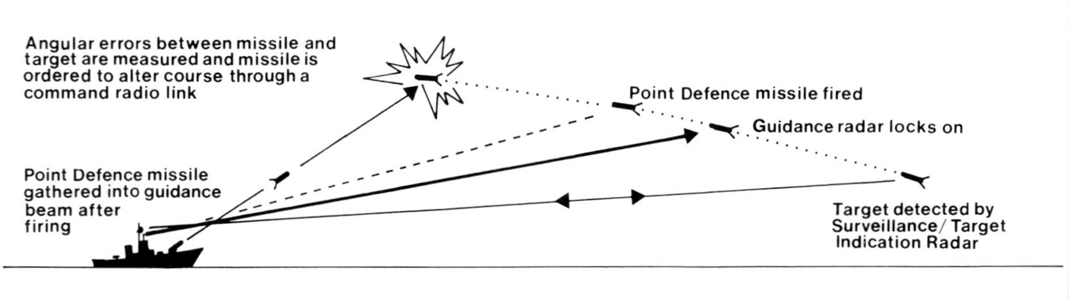

POINT DEFENCE: engaging a target with a command-to-line-of-sight system

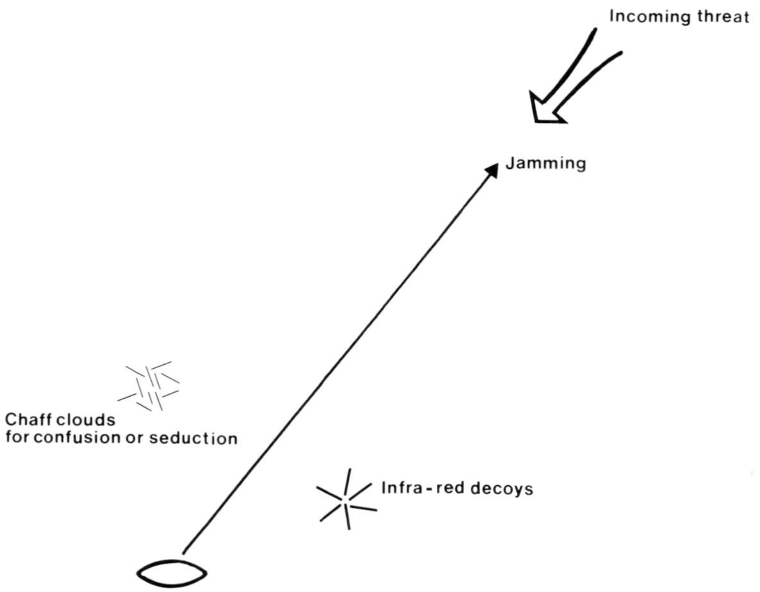

SOFT-KILL SYSTEMS

ment, is another plus. And the potential destruction of the target's delicate homing mechanisms, as well as burning lethal holes in its fabric, gives a dual possibility of neutralising it through laser attack.

The pitfalls of such a radical new system are likewise enormous. It is by no means certain that the problems of thermal blooming and attenuation, mentioned above, have been solved or are even soluble in the humid conditions of the above-sea atmosphere. Moreover, pointing more accurate than has ever been attempted is necessary to ensure that the laser beam is trained on a single spot on the incoming target; if it wavers, the energy is diffused and damage may not occur. Extensive redesign of power supplies may be needed to cope with the great surges of energy required. And finally, early equipments in particular may be of great size and weight, requiring unwelcome trade-offs in other sensors, or weapon systems, or ship's services.

To summarise briefly on hard-kill point defence systems, therefore: they are of relatively short range; they depend on rapid reaction; they require agile electronic and physical properties with a high degree of automation; low false alarm rates are very important. All this said, they do by definition generally have to cope with an easier geometrical problem — the target coming straight towards — than does the area defence system; and they can employ a greater variety of sensing systems to home them on to their targets.

So-called soft-kill systems fall into two categories: electronic and physical. The electronic systems are, broadly speaking, more or less sophisticated jammers. These transmit waveforms which, by their pattern, are calculated to confuse or divert the homing heads of incoming missiles. Clearly, the more autonomous the missile the greater are the chances of so confusing it. But accurate information on its characteristics is essential, otherwise the transmitted waveform may be ineffective or, worse, will help to attract the attacker. This information will be best obtained from ESM at the time; previously acquired intelligence can be used, but is more risky.

Physical soft-kill systems seek to divert incoming threats by presenting them with alternative targets. Clearly these must be more attractive to the attacker than the true target; thus they must be of the same kind as 'seen' by the attacker but bigger or better. For incoming missiles fitted with active radar seekers, the most popular decoy is chaff (the 'Window' of World War 2), now cut to sophisticated lengths to maximise radar response, fired from the ship at a calculated moment and blooming in bundles to represent alternative targets. Another method puts out a bundle close to the ship, which then steams rapidly away; the missile, in theory, stays locked on to the chaff and the ship escapes. For missiles fitted with infra-red seekers, infra-red decoys can be released either separately from, or combined with, the anti-radar chaff.

It is important not to discount soft-kill systems. They have had notable success in action. But it is probably clear that dependence on them for more than a proportion of the protection needed to attain the aim of air defence at sea would be foolhardy.

Air Defence Assets

Modern technology offers, then, a wide variety of assets for air defence at sea. Warning can be obtained of the approach of opposing aircraft, often at great ranges, by both active and passive means from airborne and shipborne platforms. Systems are available to track both manned aircraft and missile targets: by radar and (closer in) by optical and both active and passive infra-red means. Weapons can be deployed by air defence aircraft and by ships, the latter in area or point defence modes, with guidance and homing devices in the same spectra that are used in tracking. Finally, confusion, seduction or diversion of threats may be carried out by electromagnetic or physical means.

But all these assets cost money, often a great deal of money. Even for those states that have the resources to provide a full gamut of systems, the constantly increasing diversity, performance and sophistication of the threat will demand steady improvement of air defence assets and, perhaps even more important, their organisation and control so that their ability to achieve the aim of air defence is maximised. That organisation and control becomes even more critical in the case of those maritime forces that have only limited resources available for air defence.

In consequence, it is entirely reasonable that air defence at sea should be planned on certain principles derived from study of the technology, practical experience in exercises at sea and war-gaming ashore. It is to these principles that we now turn.

The Principles of Air Defence

Denial of Information to the Enemy

To deny information to the enemy confers several considerable advantages. First, it may present him with fewer targets, or even no targets at all. Second, it may take him longer to piece together such information as he has, thus delaying his attack. Third, it may induce him to attack subsidiary targets while the position of more important ones remains uncertain. Fourth, and conversely, it may induce him to refrain from attacking anything while attempting to locate his more important targets. Finally, it may induce him to expose his forces in a way that would not have been necessary had he been in possession of a full hand of information.

Information — and, indeed, opportunity — can of course be denied by not venturing within the surveillance range of the opposition, much less his strike range. There may be occasions when maritime forces can conduct such a policy and still be doing an effective job. In particular this may apply in low-intensity operations when the loom of a force well over the horizon is the main requirement. But when operations reach the higher level and assume the nature of war, it is likely that a larger and larger proportion of the forces involved must go in harm's way.

Thereafter, the denial of information may still be pursued, by at least three methods. The first is the strict control of electronic emission. It is almost impossible to keep a maritime force of any size at sea without making some such emission; for example, any operation of shipborne aircraft and any manoeuvring or information exchange between ships will entail use of communication nets. But the first can be extremely reticent and use airborne links, and the second can make extensive use of satellites; both are hard to detect. Radar transmissions are more subject to a need for absolute silence, particularly when radar-detection satellites such as SOSS uses are about. But here the airborne radar may be a singularly helpful asset, since it can be stationed well clear of the centre of the force and therefore its position may not give too much away. Moreover, airborne radars are to some extent directional in the vertical plane and are not the easiest of targets for a radar-detection satellite. Thus by strict control of electronic emission a force can deny a great deal of detail, if not complete knowledge of its presence, about its composition and formation.

A second means of denying information is to use random or constantly-changing formations. Here combatant forces are aided by the extended range of many modern systems, not only above water but in the field of underwater warfare. Towed anti-submarine arrays, for example, positively demand that the ships operating them stay in quiet water well away from other vessels. Area air defence systems, also, imply extended spacing between ships. Vessels are likely to be given quite large areas of sea, relatively to the force centre, in which they may manoeuvre freely. At the same time, changes in the tactical situation may dictate that these stations may be varied. The result, as viewed by the radar satellite or shadowing aircraft, is more likely than not to be ambiguous if not downright confusing; and careful control of surface forces can ensure that that effect is maximised. It should, however, be added that such dispositions are much less likely to be manageable in forces that are not fully worked up. A civilian convoy, or an ad hoc amphibious group, is not controllable in this extended way: one is likely to get a real shambles rather than a simulated one.

Finally, positive measures of deception can be employed. The out-and-out deception is typically a set-piece, like the elaborate show staged in the Pas de Calais during the D-Day landings. It is certainly not beyond the wit of any medium-sized power, much less a superpower, to mount such an operation with all the electronic and physical means now available, but it needs to be very well rehearsed or to make use of adaptable equipment and clever minds if it is to succeed for long, since it will have against it equally sophisticated means of analysis.

This brings us back to the fundamental question whether, in the modern age when the surface of the sea is so often regarded as a completely open book, with the positions of all units on it readily displayable, it is worth pursuing any notion of denying information to an opponent. It is hoped that the brief description above of how information may be denied, even partially, and the dividends that may be gained by such measures, does validate denial of information as one of the principles of air defence.

Early Warning

It has already been suggested several times in this book that events in air attack at sea tend to be very fast-moving and sudden. If a simple yardstick can be applied, it is that of the approximation to Mach 1: 600 nautical miles per hour. That is 10 miles a minute, a mile every six seconds. It is at that speed that the generality of *present-day* unpleasantnesses come towards a surface unit. Some come faster. When to that speed of approach is added the speed of any vehicle sent out to counter the attack, then one may have to start thinking in terms of 20 and 30 miles per minute or even more.

Some threats by their very nature allow the target only a short space of time in which to counter them. Typically these are sea-skimming missiles launched from submarines or small surface craft: the firing range may be as little as 10 miles, the time for detection, reaction and effective counter a minute or less. Of course, such firing ranges may mean exposure of the launch vehicle; we shall return to this shortly.

But many air threats — and particularly this applies to missiles carried by Soviet units, whether air, surface or submarine — are designed for much longer times of flight, and here early warning, both of the platform and of the missile itself, is seen at its most important. The best remedy against all missiles, and this goes for missiles like bombs and shells as well as self-propelled and self-homing varieties, is to destroy their platform before they can be launched. If early warning can give the opportunity to do this, that is the optimum condition; but even if it can only warn of incoming threats once clear of their platforms, yet in time to put defences on full alert with a maximum opportunity of acquiring, tracking and engaging them, then it will be a great deal better than a knee-jerk reaction later on.

The means of achieving early warning embrace many of the combinations of platform and sensor described earlier in this chapter. Satellite information, if available in real time, may be helpful but only to indicate ship platforms; it is doubtful whether any nation yet possesses the technology to provide timely alerts of airborne vehicles by this means. Satellite-borne passive sensors will suffer from the same limitations. Shipborne or airborne passive sensors, however, could well provide the first indication of opposing air activity and, moreover, the character of the emissions can give a guide to enemy intentions; for example, the attack mode of certain Soviet airborne radars is said to be quite different from the search mode.

However, the most commonly used means of early warning is, and will probably remain, radar. If the geographical situation is favourable, large ground-based radars may be valuable; for example, the NATO Air Defence Ground Environment (NADGE) chain along the coast of Northern Norway is placed and constructed to detect Soviet aircraft rounding the North Cape and continuing into the Norwegian Sea. A more northerly route would diminish their effective range against targets in that sea and south of the Greenland-Iceland-Faeroes Gap (where, of course, there are more ground radars).

This is a tactical advantage, but it points up a limitation of ground radars: being fixed, or at least constrained by the geography of the land, they can be circumvented. Airborne radars are not so constrained. When deployed in large, long-range, land-based aircraft they can be positioned over a considerable proportion of the earth's surface — depending, of course, on the availability and geographical distribution of bases — and modern airborne early warning (AEW) radars can detect airborne targets at ranges of up to 200 miles and handle very large numbers of them at once. They can also detect and track surface units including those that have to come close in order to fire missiles — those same missiles that cause severe air defence problems because of their short time of flight and low trajectory. Even carrier-borne AEW aircraft can now not only generate vast quantities of data on both air and ship targets, but pass it electronically to ships by means of data links and control their own side's intercept and strike

Above:
A carrier-borne early warning aircraft, the E-2C Hawkeye of the US Navy. *US Navy Official*

aircraft as well. And carrier-based aircraft have the crucial advantage that their platform is mobile.

Given the need to deny information to the enemy, the possession of AEW aircraft is an inestimable advantage, for, as was explained earlier, they do not necessarily give away the position of surface units. This does not take away entirely the need for surface ships to possess long-range air warning radars of their own. AEW may not always be available, particularly in small or detached surface forces; or it may not be positioned correctly to detect threats as they actually develop; or it may itself prove vulnerable to attack. Thus most larger surface units are fitted with powerful air warning radars. These are likely to be switched on the moment a surface force's position is known. A complete air picture will then have become more important than denial of information.

Some other means may be used to achieve early warning: acoustic arrays, bistatic radars, infra-red and low-light devices. Of these, it is probably fair to say that the first two will not, in an air defence situation, do much more than indicate that something out of the ordinary is going on in a general area perhaps some tens of miles square; while the last two are of limited range. It is also fair to say that, in such situations, no means of information gathering can safely be neglected: and this is nowhere more relevant than when considering the next principle of air defence, attack at source.

Attack at Source

Destroying or neutralising the platform that poses an air threat is the optimum way of dealing with it for several reasons. First and most obviously, it eliminates the threat from that unit. Secondly, it eliminates the re-attack potential of that unit. Thirdly, it reduces the opponent's trained manpower and may adversely affect his morale. Finally, it may preoccupy him in protecting his remaining platforms and thereby draw off resources which otherwise could be used to attack one's own.

This principle should, whenever circumstances — including political circumstances — allow, be extended to attacking the land bases from which air threats operate. It is well known, for example, that the Kola Inlet complex contains the main stores, workshops, forward repair yards, communication centres and airfields for the Soviet Northern Fleet and Northern Fleet Air Force, and the view has often been expressed in United States Navy circles that early attack on this complex would be a necessary element of a forward move by the US Strike Fleet into the Norwegian Sea.

On the other hand, there will be many operations in which this option is not available. The British campaign in the South Atlantic in 1982 is a case in point: political considerations precluded attacks on Argentinian mainland airfields. Similarly, the first three USN operations off the Libyan coast in 1986 did not include any attacks on airfields or, indeed, any shore complex except an air defence radar; the attack on shore that finally was made was apparently punitive in intent and not anything to do with the air defence of the US Sixth Fleet.

Moreover, this kind of attack at source cannot be thought of as a panacea, even in operational

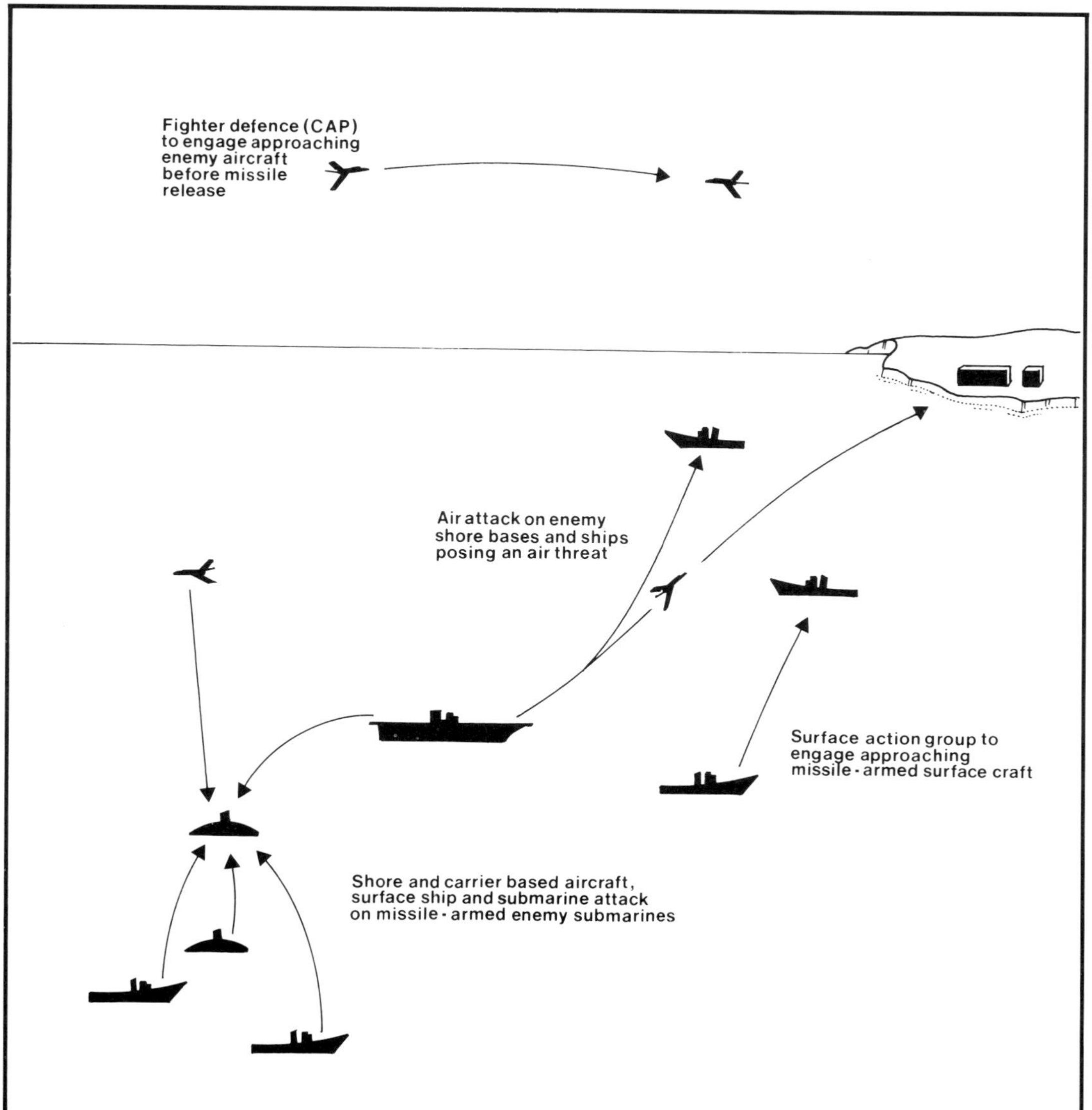

ATTACK AT SOURCE: some examples

terms. Countermeasures available to an opponent are numerous. Airfields can be and frequently are repaired, in spite of modern runway cratering weaponry; stores can be dispersed or salvaged; hardened sites make attack problems more difficult; advanced and improvised bases can be put into operation. Thus, though carrying the air defence forward to its ultimate limit is a good principle and always deserves consideration, the factors need careful weighing when resources are being planned and allocated.

It is a fair assumption that, in spite of any such action that may be feasible, the opponent will have units at sea and in the air that continue to pose an air threat. Action against these units is therefore part of the attack-at-source principle.

Submarines armed with anti-ship missiles may be treated as an underwater problem, and the reader is referred to the author's *Anti-Submarine Warfare* in this series for a discussion of that. Those submarines that have to surface in order to fire their missiles, and may have to stay on the surface to direct them in mid-course, will be vulnerable to detection by airborne radars, and to air attack if it can be mounted in time: but reaction would need to be very swift in such a case if the missiles were to be rendered ineffective.

Surface units posing an air threat — which can be taken to mean a threat either by manned aircraft or missile — can be struck by a variety of means. Submarines may be deployed specifically against them (it is worth recalling that in the Battle of the Philippine Sea a critically important early strike was made by a US submarine) and can now use missile or torpedo armament to attack; surface action groups may be formed to deal with them; or they may be subjected to air attack by missile or directly by aircraft both carrier and shore-based. The number of options is indeed formidable, but the one thing one cannot do with surface threats is to ignore them. Moreover, they are sometimes a good deal more difficult to find than might be thought. They will be following the Principles of Air Defence too.

The third type of platform to be subjected to attack-at-source is the air platform, and it sometimes seems to bulk so large in the calculations of air defence practitioners that it diverts attention from the other two. This introduces a dangerous imbalance that must be resisted. Nevertheless, it is true that the air platform moves much faster than any other, and so when a threat develops it is indeed, as the jargon has it, time-urgent. Therefore the provision made to meet it and to impose attrition on the incoming aircraft, has to be sufficient in scale, precise in conception and swift in operation.

The principal problem — and it is at its toughest against the Soviet Naval Air Force with its long-range air-to-surface missiles — is to get at the parent aircraft before missile release. This entails intercepting them at ranges well over a hundred miles from their target; a glance at the tables in Chapter 1 will indicate the scale of the problem. But that is only the start of it. Which is the target? If it could be any ship in a dispersed force, and it is desired to protect all those ships, the intercept point is advanced. Then consideration must be given to the time needed to get the intercepting force out there; is it to fly from shore, or from a carrier deck, or will it already be on Combat Air Patrol (CAP)? If a CAP is to be mounted, how far up-threat can it prudently be stationed? Interceptors may need to be controlled by a unit with good tracking and processing facilities (their parent ship, a radar picket ship or an AEW aircraft); can that unit be stationed to achieve a timely interception?

When these considerations are put together and given numerical values, they often seem to add up to a demand for early information on the enemy that cannot be fulfilled. The problems can be represented in geographical plan, as in Figure 12, or as an attack diagram as in Figure 13. However, it should be remembered that the Soviet Naval Air Force in all its Northern and Pacific Fleet panoply does represent a worst operational case, and the assets (including early warning assets) that can be deployed against it are considerable, and in some aspects unique, in their scope. For medium-power threats, with missiles whose range is measured in scores of miles and with limited numbers of aircraft, the problem, though still acute, is more manageable.

Attack at source, then, is a most important component of air defence at sea. If completely successful it can fully protect surface forces and achieve the aim. But even the United States Navy, which certainly aspires to such a result, is conscious that total attrition of the platforms capable of air attack is not the most likely outcome. In USN language, 'leakers' can be expected. Sceptics may expect the leaks to be a good deal more extensive than the technologically confident USN claims. For other nations and navies, the aspiration is clearly much less realistic still. Therefore, for all, there must be a fourth principle of air defence at sea: layered defence against missile attack.

Layered Defence

In essence, air defence at sea is 'layered' from the point of origin of the threat to its intended point of impact. Attack at source is part of the process, the outer layers if you like; the principles are complementary, not separate. But this section will deal mainly with defence against anti-ship missiles once they have been fired — in other words, if attack-at-source has not been fully effective.

The sensors and weapons that may be available to a surface force have already been described in broad terms. They will include radars for warning, acquisition and guidance; passive electromagnetic sensors; infra-red, low-light and laser sensors; fighter aircraft, hard-kill missiles and guns; and soft-kill equipments such as jammers and chaff.

They will add up to a list of air defence assets, and on the sufficiency, deployment and employment of these assets depends the success of the air defence operation. The necessary decisions devolve on the force commander, and will require all his experience and skill; this is why there are admirals at sea. He will not only have air defence to think about. The overall mission of the force must always be kept in mind and may influence his dispositions; so will the anti-submarine situation.

The first assessment to be made is on the nature of the threat. Clearly a threat such as is posed by the Soviet Northern Fleet Air Force, with predominantly long-range air-to-surface missiles, will demand particular kinds of disposition and organisation, widely different from those required solely against close-range sea-skimming missiles, fired from surface craft or submarines. Multiple threats, including all these categories and some others, will need skilful compromise in the defensive arrangements.

The direction of the threat, sometimes called the threat axis, may be used as a planning tool. There are situations, particularly when the main threat is from aircraft working from known shore bases, where it is right to bias defence assets heavily in the expected direction of attack. But the danger of being outflanked, either by mobile units such as surface craft or submarines, or by aircraft showing an unexpected turn of range, has always to be borne in mind.

Having made his threat assessment as precise as he prudently can, the force commander must now dispose his assets in the best way to meet it. He will be particularly concerned to protect his high value units, the operational core of his force; these will certainly include aircraft carriers if he has them, and probably the principal replenishment units without which the force cannot stay in fighting

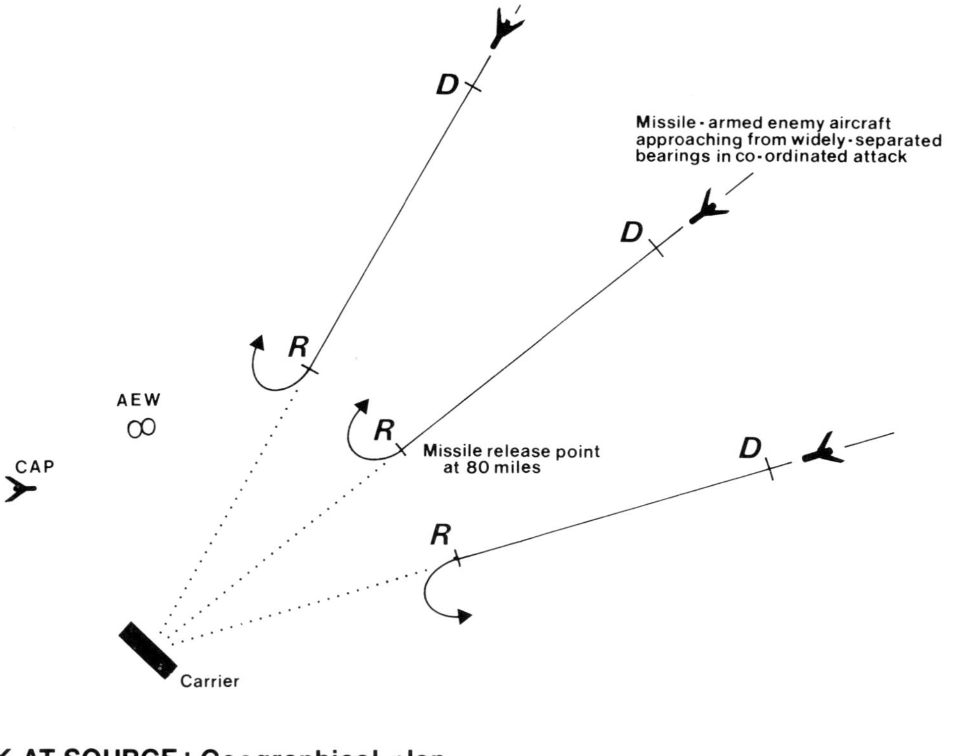

ATTACK AT SOURCE: Geographical plan

In this convergent attack by missile-armed aircraft, AEW detection (*D*) is taken to be 120 miles. The CAP, directed by AEW, can readily intercept the northern-most raid before the point of missile release (*R*). It can barely intercept the middle one, and cannot intercept the eastern-most.

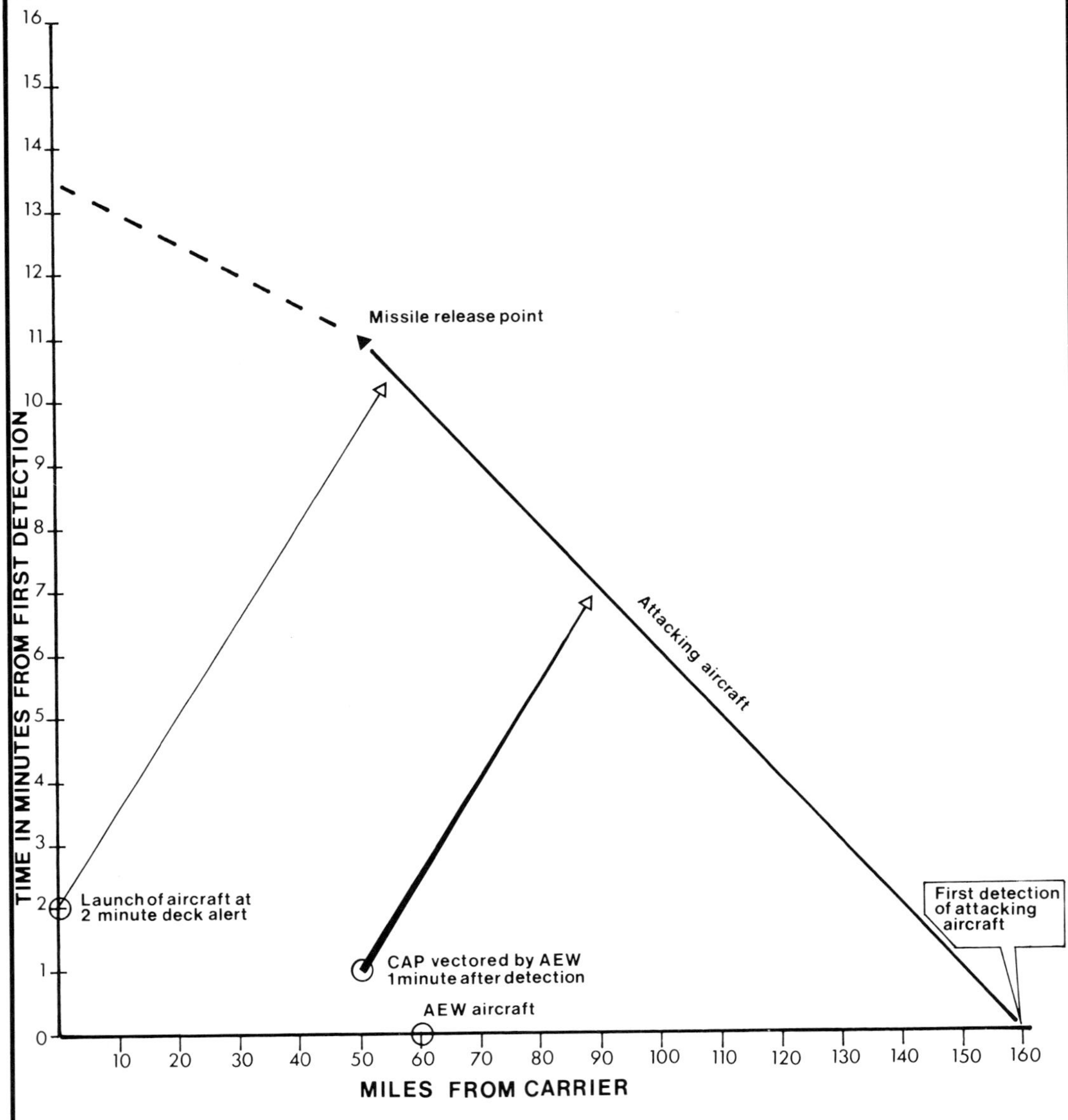

ATTACK AT SOURCE : use of the attack diagram

Attacking aircraft approach at 600kt. Missile release point is 50 miles from target – missiles fly at 1,200kt. AEW aircraft stationed 60 miles up threat vectors CAP from station 50 miles up threat to intercept at 500kt. Intercept is readily achieved by CAP, but with much more difficulty by aircraft at 2 minutes deck alert, before attacking aircraft release their missiles.

(Diagram after D J Martin, Thorn EMI Electronics)

condition for long, and also major amphibious units if the force is of that nature.

If he has AEW aircraft, they are best stationed up-threat; this helps the attack-at-source principle too. Other surveillance aircraft, notably those on anti-submarine patrol, may be valuable not only in warning of developing submarine-based missile threats but of surface units as well. Any aircraft may — though this will be occasional and fortuitous — detect the firing of missiles by an enemy unit, whether this unit has previously been detected or not.

In general, however, the approach of missiles is more likely to be first detected by ships themselves. Whether the means are active or passive depends on the emission control policy in force at the time — here is another way in which the principles of air defence interact. Stationing a ship, particularly one fitted with good radar and ESM, up-threat from the force may improve warning time; if that ship also has area defence missiles it may be well placed to form an outer layer of anti-missile defence. But such pickets are isolated and vulnerable, and not only to air attack: submarines particularly may fancy their chances against them. Some mutual support, whether from a towed-array unit or other means, may have to be arranged.

Closer in to the high value units the layering can be made thicker, and indeed needs to be, since it has to cope not only with long-range missiles that have penetrated the outer layer but with short-range missiles fired from inside it. Here the emphasis shifts from area defence to point defence weapons. The former are still valuable, particularly so if high value units themselves are not fitted with them; for example, a USN aircraft carrier, even the largest, will need area-defence cruisers and destroyers stationed relatively close at hand for its protection. But point defence weapons will be the principal means of self-defence, and the shorter the firing range of enemy missiles the greater the emphasis on this sort of defence will be. There is one variation on this theme, the stationing of ships with good point-defence systems very close to inadequately defended high value units as 'minders'. This is a device arising out of the material deficiencies of particular units and is not an ideal practice.

Finally, the value and potential of soft-kill systems may well be taken into account in the disposition of the layers. Ships with particularly effective jammers or the most modern chaff systems may, if other circumstances permit, be put in the more vulnerable positions. This minor example makes a general point: any naval force will offer a diversity of assets in a unique combination, and only careful balancing of the factors will achieve the optimum result.

But the best of dispositions will be of no avail if

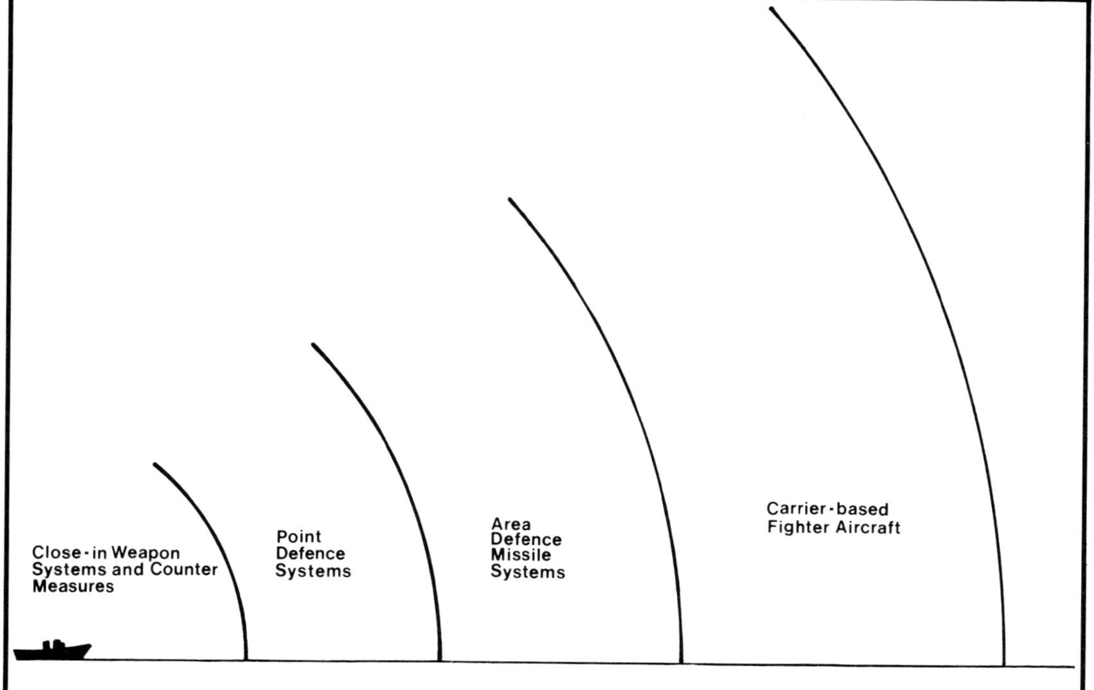

LAYERED DEFENCE

NOTE: 1 The diagram is greatly simplified. In practice the layers overlap even for a single ship, and very much more so for a balanced force.

2 In the zone beyond carrier-based fighters, shorebased fighters (assuming favourably placed bases) may form an additional layer.

the handling of the force's assets, when threat materialises, is inadequate or slipshod. There are two main requirements here. The first is a comprehensive system of real-time exchange of data and instructions. Modern electronic technology has made great strides in giving navies the ability to acquire, process and present great quantities of information, and to pass orders quickly and securely both between and within ships and aircraft. There are pitfalls in this wonderland of chips. The very profusion of information means careful sorting of the vital, the important, the interesting and the irrelevant. The very precision of command systems may tempt superiors into ordering too much and delegating too little.

All this leads simply to a statement of the second requirement: training. All the personnel concerned in the air defence scene must know their own jobs, and these will often entail very quick-reaction drills in carefully-rehearsed contingencies. For example, the detection of an incoming sea-skimming missile has to set in train a very rapid set of actions by radar operators, weapon controllers, ECM operators and chaff operators, all under the supervision of the command or its delegated representative. That action must be possible 24hr a day in an operational area, so, since men must sleep sometime, more than one team must be capable of doing it. A good deal of cross-training is therefore required; but over and above this, it is necessary that all those concerned in the air defence scene should have an idea of the contribution they are making to the whole, otherwise they will not be able to do their part with the necessary perception.

Avoidance of Blue-on-Blue

Nowhere is the need for perception, and understanding of the totality of the air defence position, more apparent than in the requirement of avoiding attacks on one's own forces. This, in the view of this writer, is certainly worth elevating into a Principle. The history of war, and of war at sea no less than on land, is full of such instances of 'blue-on-blue'. In the very rapidly moving air defence scene, the hazards are obvious.

The most difficult problem is probably that of safeguarding one's own aircraft, and it falls into two distinct parts: aircraft originating from one's own force, and those coming from remote bases.

Aircraft based within the force may themselves be air defence aircraft, or they may be strike aircraft; in both these cases they will be fast, fixed-wing machines. And a naval force will almost certainly have many helicopters operating not only from carriers but from smaller warships and, it may be, auxiliaries.

There are two basic ways of seeking to ensure that these aircraft are not attacked by one's own force. The first is to designate sectors or zones in which the use of anti-aircraft and anti-missile systems is restricted because friendly aircraft are present or expected to be present in them. The restriction can be extremely tight (fire may be opened only if own ship is under direct attack), slightly less so (fire may be opened on units positively identified as hostile) or relatively relaxed (fire may be opened on unidentified aircraft).

Below:
A rudimentary identification device: but whatever the means 'blue-on-blue' must be avoided. *Crown Copyright*

These procedures — not to be confused with Rules of Engagement, which are covered elsewhere — are suitable for forces where not all the units are in possession of a full air picture.

More precise, and less operationally restricting, is the provision to all units in the force of accurate information on the position of all friendly aircraft operating round it. In these days of automatic data exchange this is increasingly possible, even if only a few units are operating primary sensors such as radar. So long as the information gets around inside each ship in the force, weapon controllers should have no difficulty in avoiding attacks on their own side even in complex situations.

This second method is seen to be even more desirable when one considers the safeguarding of aircraft coming into the force's ambit from remote bases, whether on land or from other forces at sea. Such aircraft may be under the control of agencies other than the force commander, and their movements less predictable therefore. Their general direction of approach may well be known, and at that stage sector procedures might be used, but as soon as the aircraft can be detected and identified they must be tracked and their positions frequently updated in the whole force.

Identification Friend or Foe (IFF) systems should, in theory, greatly simplify the blue-on-blue problem. The reception of positive coded electronic responses from all friendly aircraft ought to deter weapon controllers from opening fire on them. However, it is clear from the open press that great difficulty has been experienced, in NATO at any rate, in evolving an unjammable, uncrackable, reliable IFF system; and the dangers of relying on an IFF system that is less than that are so evident as not to need elaboration.

A Pattern of Operations

A pattern of air defence operations at sea emerges from this discussion and development of the principles. It is not universal; there may be many operational situations where it is necessarily pulled out of shape or lacks many of the basic elements. But it may be useful to outline it here, to put in context the description of weapon systems in Chapter 3 and of some specific types of air defence battle in Chapter 4.

A maritime force (and that term may be taken to include any group of ships which it is desired to protect against air attack) may well begin its career without being under immediate threat. It may come under threat because conflict starts — in which case it will have a difficult transition time in which Rules of Engagement will figure prominently, and the identification of hostile intent will be of extraordinary importance — or because, already at war, it enters a geographical area where opposing forces are operating.

In both these forms of transition, the denial of information to the enemy is likely to be an important element. Even if, for politico-military reasons, it is desired that the presence of the force should be known, the details of its operations should be concealed as far as possible. Denial however, because it involves restricted emission policies, militates against the gaining of early warning; here AEW aircraft can make a most important contribution.

Once air threats develop, the denial of information becomes largely a matter of confusing the opposition by manoeuvre and, perhaps, deception. But the force must use more resources to gain warning of impending attacks and to control the means of repelling them. It may still restrict its emissions and make maximum use of remote sensors such as AEW, other patrolling aircraft, and space and bottom-acoustic systems.

These last may be particularly valuable in detecting threats from mobile platforms such as submarines and surface ships, give the opportunity for relocation by more local means (air probe, or the ubiquitous AEW or patrol aircraft) and allow attack-at-source counters to be developed by air strike, surface action group or fleet submarines.

In attacking the airborne missile threat at source, extensive use will be made of Combat Air Patrols, generally directed by AEW aircraft; both they and the CAPs themselves will be stationed up-threat if a threat axis can be identified. CAPs may be provided not only from ship platforms but from shore bases if these are within range of the force.

To combat missiles that, in spite of attack at source, are fired at the force, a system of layered anti-missile defence will be employed. The outer layer will consist of area air defence systems, the inner layers point defence systems including missiles, guns, jammers and chaff decoys. The closer the range at which such systems are effective, the greater is the need for rapid means of, and procedures for, bringing them into action; but the whole layered-defence system depends on excellent training, rapid data exchange and clear direction and co-ordination by the command at all levels.

This requirement too is evident in the need to avoid any attack on friendly forces. This is done principally by ensuring that every unit in the force is fully informed of the whereabouts of all friendly airborne units; but such comprehensive information can be backed by procedures whereby the firing of weaponry is restricted in certain zones or sectors around or within the force.

It will be a well-endowed maritime force that can do all this. The next chapter reviews what is available — at a price — in the way of air defence resources today.

Components of layered air defence

Left:
Fighter aircraft: a F-14 comes aboard USS *Carl Vinson*. *Frank Mormillo*

Below:
Airborne Early Warning: RN Sea King helicopters modified to take Searchwater radar.
Crown Copyright

Bottom:
Area air defence systems: a Standard missile fired by USS *Ticonderoga*.
US Navy Official

Above left:
Point defence missiles: Sea Wolf.
British Aerospace Dynamics Group

Above right:
Guns: the Breda Meccanica twin mounting.
Studio Grafico Restani

Right:
Close-in multi-barrelled guns: Vulcan Phalanx.
Crown Copyright

Right:
Chaff launchers: the Shield system.
Plessey Aerospace

Below:
Clear and rapid data processing and exchange.
Ferranti Computer Systems Ltd

CHAPTER 3

The Weapons

The range of equipments available for air defence at sea is extremely wide, as the briefest of scans of the text (and particularly the advertisements) of any modern defence magazine will indicate. This chapter will necessarily have to be selective in the equipments it discusses, and if it concentrates on those available to NATO countries that is because the open society of the West supplies more details than others do.

It must, moreover, often treat sensors and weapons as part of an integrated whole, because that is what the systems themselves do. Where a sensor or other equipment does duty in more than one field, its details will be covered only once; in other places there will be the merest cross-reference.

There is one further self-imposed limitation in this chapter. It will not go into the business of air combat, nor of surface action, nor of anti-submarine warfare. These are all implementations of the attack-at-source principle, certainly; but they belong properly to other techniques of warfare, and a line must be drawn somewhere.

Air Systems

Airborne Early Warning (AEW)

The United States was first into the field of Airborne Early Warning; even at the end of World War 2, there were aircraft flying which could transmit, to remote displays in ships, a radar picture of the surrounding sea out to over 100 miles. The American pre-eminence has been maintained ever since.

The main shore-based AEW aircraft now operated by the USA (and by NATO) is the familiar E-3 Sentry, whose equally familiar acronym AWACS (Airborne Warning and Control System) describes its function accurately. On a sturdy four-engined Boeing airframe, its APY-1 radar and the associated processors have the ability to handle hundreds of air targets at a time, and its 13-operator crew can control multiple fighter operations. The E-3 was designed primarily as an overland AEW aircraft; its radar makes extensive use of doppler to identify fast-moving targets and reject slow-moving or stationary ones. It is probably fair to say, therefore, that it is not optimised for sea surveillance, which of course includes ship as well as air targets.

The US aircraft that is optimised for such a task is the Grumman E-2C Hawkeye, which operates from US fixed-wing carriers. It operates at heights of between 20,000ft and 30,000ft and has an endurance of up to six hours. With its APS-125 radar feeding an onboard computer, its crew of three can control, so it is said, 25 simultaneous intercepts. There is no doubt about the confidence placed by the USN in this aircraft. Many accounts stress the primacy of its role not only in warning of both sea and air targets, but in the control of other aircraft, notably interceptors.

No other nation remotely approaches this sort of capacity. Britain aspired to it in the Nimrod AEW3, but that aspiration finally came to nothing. The British decision to reject AWACS and rely on an all-British development was made in 1977; the unanimous approval of the House of Commons, almost unique in the writer's memory, seemed at the time to be the worst of auguries. This turned out, alas, to be all too true. A difficult radar concept was matched with a slow computer of limited capacity; development was plagued with all conceivable technical problems of which, by authoritative accounts, the most prevalent were fluctuating transmitter power, too many radar sidelobes and consequent unexpected clutter, and over-sensitive receivers, all of them compounding the problems of the already inadequate computer and causing massive overload and failure. Changes in the Staff Requirement during development did nothing to help. The AWACS has now been chosen instead. It should certainly fill the oversea airborne early warning and fighter control role admirably, when operated in that role.

But AWACS is a large shore-based aircraft. The Falklands conflict proved again (as if proof were needed) that ships beyond the range of shore bases need organic AEW if their air defence is to be effective, and in consequence the Royal Navy has developed an AEW helicopter in the Sea King Mk 6. The radar is an adaptation of the successful Searchwater fitted in the Nimrod 2 ASW aircraft, and though it is a far cry from the high sophistication of US systems it does provide the fleet with warning, and limited control, in the face of air threats.

The French carrier-borne Alize, a fixed-wing turbo-prob aircraft brought into service some 20 years ago for ASW duties, is said now to have an AEW role after being retrofitted with Iguane radar. In view of its endurance, service ceiling and primary ASW role, this capability must be limited.

The Soviet Union has for nearly two decades operated Tu-126 'Moss' AEW aircraft, but these have been mainly in the overland environment and have in any case nothing like the capacity and sophistication of their American counterparts. Reports suggest that the Il-76 'Candid' is emerging as a successor in this role. There is much ground to be made up. The oversea environment appears even less well catered for; the 'Bear' is of course an excellent reconnaissance aircraft against surface targets, but this does not extend to air warning, and no form of ship-based AEW has been observed. The question will become more acute, for the Soviet Union could not logically plan to send a large fixed-wing aircraft carrier to sea without an adequate AEW development on the horizon.

Air Defence Aircraft

In carrier-based aircraft the United States Navy is pre-eminent in a way that now applies to hardly any other set of weapon systems. Generation has succeeded generation with steady increases in performance and capacity. The current inventory of air defence aircraft is based upon the F-14 Tomcat, an extremely sophisticated twin-engined, two-seat aircraft with an onboard radar fully capable of acquiring targets in time for its long-range Phoenix air-to-air missiles to take maximum effect; and the F-18 Hornet, a single-seat aircraft, lighter than the F-14, but still immensely capable in the air defence as well as attack roles, with medium and short-range air-to-air missiles and an onboard computer for their control.

The endurance of these aircraft is quoted as three hours or slightly less. Aircraft endurance is always a consideration in air operations, but in carrier operations it is absolutely critical. The whole pattern of a carrier's manoeuvres is based upon the flying programme, and this in turn must take account of the amount of time each type of aircraft, on each specific mission, can safely stay in the sky. The tidiest way of constructing a flying programme, particularly if a CAP is to be maintained, is to work a 'cycle' based on the endurance of the CAP aircraft. Thus it might be that a US carrier working a CAP of F-18s would be on a basic cycle of 2hr 10min, with AEW and ASW aircraft working a double cycle of 4hr 20min. In this way the carrier needs only to turn into wind for flying at the intervals prescribed by the cycle, and can go in the direction required by its mission

Below:
The US Navy's most capable ship-based fighter aircraft, the F-14A Tomcat, is here seen carrying full AAW armament. *US Navy Official by PH1 William A. Shayka*

for the majority of the time. Helicopter operations, which from a carrier deck do not necessarily need into-wind courses, can be interspersed.

That is the solution at its simplest. Even at the planning stage complications arise: what, for example, of the situation where CAPs are stationed some hundreds of miles from the carrier and yet have to be relieved on station because of an acute air threat? In the time between flying off the relief CAP and the recovery of the relieved one, has the carrier time to turn back out of wind? Does it need to? Such problems crowd in.

In practice, the best-laid flying programme can be in for nasty surprises. It can be disrupted by accident, weather or other natural contingencies that make aircraft late or early; and it can be disrupted by the enemy, who ideally would always like to attack just as a carrier is about to turn into wind. It can also be disrupted by changes of plan by the command; and these may not be simply an irritant but literally fatal, as was Nagumo's decision to rearm his strike aircraft at Midway.

Similar practices are adopted by all navies operating fixed-wing, conventional take-off aircraft from carriers. The air defence aircraft available to the smaller carrier forces are limited in both numbers and capability: the French have F-8E Crusaders, an American design armed with French medium-range M530 air-to-air missiles; the Argentine Navy has A-4 Skyhawk aircraft operating from its single carrier.

In the medium-sized navies there is a distinct shift towards the employment of VSTOL aircraft for air defence. This was led by the United Kingdom when it adopted, not without reluctance on the part of many, the Sea Harrier in the early 1970s. Initially the Sea Harrier was regarded as an aircraft to harass shadowers, conduct surface probes and pose the threat of limited strike against ships. It emerged — well before the Falklands conflict, in fact — as an air defence aircraft in its own right, scoring notable successes in fully monitored exercises against aircraft as sophisticated as the F-15. In the South Atlantic it proved itself. Harriers shot down over 20 Argentine aircraft and not one was lost in air combat. This was due partly to the agility of the aircraft, partly to the excellent training and spirit of the pilots, and partly to the reliable Sidewinder AIM-9L short-range, heat-seeking missile. Some limitations in the fighter role emerged, particularly in the Blue Fox radar fitted in the aircraft; this is now being replaced by the Blue Vixen, which has the ability to look down.

As an air defence aircraft to be operated from a carrier, the Harrier — and, by extension, all VSTOL aircraft — has some plus and some minus points. Its plus points are its intrinsic flexibility not only in the air but from the deck, in that it does not necessarily need a turn into wind for landing on; its very high serviceability; its relatively small size. Its main minus point is its short endurance, even with skijump take-off which so much enhances its ability to carry a useful load. It cannot sustain CAP stations a long way from its parent carrier, and the flying cycle is in some ways an even harder taskmaster than it is in a conventional take-off and landing (CTOL) carrier.

When all is said and done, though, the VSTOL aircraft is a satisfactory solution to a requirement for an outer layer of air defence against a limited threat, particularly if it can call upon AEW for alerting and control. It is, therefore, no wonder that several navies which previously either operated limited CTOL carriers, or none at all, have now opted for a VSTOL solution. India operates Harriers from its single carrier *Vikrant*, to be replaced by the newly-purchased *Hermes* (*Virat*); Spain does likewise from the newly-commissioned *Principe de Asturias*. Italy and Brazil may yet follow suit. Both have carriers easily adaptable to the VSTOL role.

This leads, inevitably, to discussion of the Soviet fighter air defence capability. The Yak-38 'Forger' is so far the only carrier-borne Soviet fighter. VSTOL in operation, it displays all the limitations of the genre without, apparently, many of its virtues. There is nothing in the evidence to suggest that it has the Sea Harrier's all-weather flexibility, agility or fighting qualities.

Where the Soviets go from here is a difficult guess. Plans for their large new nuclear-powered CTOL carriers must surely have advanced to the stage where the first-generation air defence aircraft has been chosen; that will be an aircraft as close to F-15 capability as they can manage. It will, as has already been said, need AEW aircraft to give full effectiveness to the system. These are two giant strides they have to take. But they also have a very large investment in the 'Kiev' class, where so far the inadequate 'Forger' is the only air defence aircraft. Will they also try to develop a fully-capable VSTOL fighter — another massive project? Some commentators argue that with the new Soviet leadership, it will be impossible for the Soviet Navy to get sufficient priority for all these things. That may turn out to be so, but they have done quite well in the priority struggle for the last 30 years.

Shore-Based Air Defence

As has already been pointed out, shore-based AEW can often be an important component of warning. But in many geographical circumstances the shore can provide other systems that greatly enhance the air defence of maritime units.

The capacity of the shore to accommodate very

large and powerful air warning radars is nowhere more apparent than in the NATO Air Defence Ground Environment (NADGE) chain round the coast of northern Norway, on the Faeroes, in Iceland and in the United Kingdom. Because the radars, whatever their power, are limited by the radar horizon, the cover of the Norwegian Sea is not complete, but NADGE is extremely significant not only in its direct ability to contribute to the total air picture, but also to inhibit the movements of Soviet aircraft that wish, for whatever reason, to conceal their presence.

It is of course necessary to integrate the information from such radars with all the rest of the data available to maritime units, and it must always be remembered that their purpose covers warning to the shore as well as the sea. Thus extensive shore-side processing and sifting, plus good communication links with forces at sea, are necessary in order to maximise the system's value.

Below:
The Soviet Navy so far has only limited-capability VSTOL carriers such as *Kiev* to provide organic fighter defence at sea.
Crown Copyright by David O'Brien

SHIPBORNE FIGHTER AIRCRAFT

Type	*Country of origin*	*Length (ft/m)*	*Wingspan (ft/m)*	*Speed (Mach)*	*Engines*	*Endurance*	*Armament*	*Remarks*
F-14A/D Tomcat	USA	62.7/ 19.1	64.1/ 19.53	2.34	2×Pratt & Whitney TF-30-P-414A	3 hours	2 Phoenix LRAAM 2 Sidewinder SRAAM 1 20mm gun	—
F/A-18A Hornet	USA	56.0/ 17.07	37.5/ 11.43	1.8	2×General Electric F-404-GE-400	2 hours +	4 Sparrow MRAAM 2 Sidewinder SRAAM 1 20mm cannon	Single-seat
Sea Harrier FRS1	UK	47.6/ 14.5	25.26/ 7.7	0.96	1×Rolls-Royce Pegasus 194 vectored thrust	1 hour +	4 Sidewinder SRAAM 1 20mm cannon	VSTOL. Also India, Spain
F-8E(FN)	USA	54.5/ 16.61	35.17/ 10.72	1.8	1×Pratt & Whitney J-57 P-20A	2½ hours	4 20mm guns M530 SRAAM	French Navy
Yak-38 'Forger'	USSR	49.2/ 15.0	24.0/ 7.32	1.1	1×turbojet 2×lift jets	1 hour max	2 23mm cannon	VSTOL. No radar
A-4Q Skyhawk	USA	40.26/ 12.27	27.5/ 8.38	0.9	1×Pratt & Whitney J-52P-408A	2 hours	Cannon	Argentine Navy

Steady progress appears to have been made on this in recent years.

The Norwegian Sea presents the archetype of shore-based air warning systems, but similar chains exist in the Baltic Approaches and Mediterranean where the geography generates more intricate correlation and co-ordination problems. It is also worth pointing out that bistatic or over-the-horizon radars (OTHR) have been mentioned in the Western press for years and one 'Western' country, Australia, has recently put much money and some trust in the Jindalee OTHR project. These shore-based radars can be expected to alert their owners to unusual air activity at great ranges; whether they can do much more is uncertain. On their side the Russians have a very comprehensive system of coast-based air warning radars, with a primary mission of detecting penetrations of Soviet airspace. How closely this can be integrated for air defence purposes with the information available to ships at sea is uncertain. The Soviet forces are in doctrine dedicated to the notion of co-ordinated operations and information, and it must therefore be their aim to provide all available data to units at sea with a need for air defence.

Fighter aircraft operating from shore can also, within the limits of geography and of their performance, be a powerful aid to air defence at sea. The United States Navy, because of its comprehensive sea-based air power, puts no reliance on such forces. The British system is somewhat different. No front-line shore-based air defence aircraft are owned by the Navy, but the Royal Air Force's air defence Tornado F2s/F3s have a mission of maritime air defence and a radius of action, particularly if air-to-air refuelling is employed, over a large part of the Norwegian Sea and northeastern Atlantic. Their interdictor/strike counterparts could have an attack-at-source role, and in this could be joined by the RAF's Buccaneer aircraft.

The Soviet Navy has about 70 shore-based air defence aircraft split equally between the Baltic and Black Sea Fleets. They are Su-17 'Fitters', a far from modern design and suitable for local air defence or, at most, covering short-range amphibious operations. It is possible that in a war situation, help might be available from the PVO, the very large air force held for the air defence of the Soviet Union, but its effectiveness might be limited owing to its lack of training in oversea operations.

This last point leads to a general observation that must close this section. Shore-based aircraft do have a serious part to play in air defence at sea, particularly as so many critical naval operations take place not too far from shore. But to be effective, those shore-based aircraft must have trained repeatedly in the maritime role and correct arrangements must have been made well in advance for their command and control. To take an example, aircraft likely to operate in support of US carrier groups must be accustomed to working under the control of AEW aircraft; if they are unable to do so, it will be difficult indeed to integrate them into the air defence of the group.

Ship Systems

Warning Sensors

Long-range air warning radars available to ships now fall into several categories. The kind which has become most familiar over the years, and is still by far the most commonly found, operates on relatively low frequency and has a rotating aerial designed to throw out a beam which is narrow — only a few degrees — in the horizontal plane. In the vertical plane the purpose of the aerial design is quite the reverse — that is, to make cover as gapless as possible in spite of fluctuations caused by differing raypaths.

But if the basic concept of these conventional air warning radars has remained unchanged for years, their detailed design certainly has not. The drive to improve performance has concentrated on suppression of unwanted clutter, focusing of the beam in the desired direction and eliminating sidelobes, and making life as difficult as possible for the jammer. Thus radars are now offered with banks of filters and onward processors, commonly using pulse compression, doppler and fast fourier transform techniques; with stabilised and very carefully-designed aerials; and with frequency agility that, by randomly varying the frequency either from pulse to pulse or from one bunch of pulses to another, makes it impossible for a jammer to concentrate its power effectively.

Typical of such radars are the American SPS-49, widely fitted in USN carriers and smaller ships; the Dutch LW-08 and DA-08, found in many European and some South American navies; the French DRBV 26 (Jupiter) standard air warning radar of the French Navy, and the smaller and lighter Sea Tiger to be fitted in the French-built Saudi Arabian frigates; the Italian RAN-20; and the Soviet series with NATO codenames 'Head Net' and 'Big Net'. The Royal Navy employs the Type 1022, Dutch LW-08 equipment with a Marconi aerial.

A second category of air warning radars employs a rotating aerial to give a narrow beam in azimuth, but in addition has electronic scanning in the vertical plane so that the elevation (and therefore height) of a target can be determined. The operational advantages of such a radar, particularly its ability to indicate quickly to the pencil beam of a missile guidance radar the area of sky it must search to acquire its target, need no stressing;

Left:
An air warning radar of conventional design, the Jupiter-II. *Thomson-CSF*

Below left:
Even small craft can be given a limited air warning capability: this is the Triton-S surveillance radar. *Thomson-CSF*

the penalty is that to complete even an electronic scan, time is needed and either data rate or peak power on target must be reduced.

Radars of this kind include the Soviet 'Top Sail'/'Top Plate'/'Top Steer' series; the American SPS-48 series; the Dutch '3-D' radar fitted in the 'Tromp' class; and the French DRB-10, though this scans physically in the vertical plane and must have considerable limitations. The Royal Navy has no system capable of height-finding in this way.

The latest developments in air warning radars employ electronic scanning in both horizontal and vertical planes. In these phased-array radars, the aerial does not rotate physically. It consists essentially of a multitude of phase shifters each of which, by virtue of its position and the electronc inputs to it, delivers (when triggered) a beam in a certain direction relative to the ship. By the use of digital processing techniques, the received responses can be analysed automatically and the transmitter instantly programmed to give more 'dwell time' to those parts of the aerial which are looking at targets as opposed to clutter. In addition to making aerials adaptive in this way, the designer can, without too much extra difficulty, incorporate frequency agility over a wide band and anti-clutter techniques.

The biggest difficulty is money; such radars are quite as complicated as their description implies, and up to the present time have required a very large research, development and production base. Another difficulty may be topweight; so far, again, the required aerial arrays are large and heavy and though they do not necessarily need a top-of-the-mast position they do impose constructional constraints.

The United States was not deterred by these considerations and has become the acknowledged world leader in phased array radars. The classic ship version is the AEGIS AN/SPY-1 in the 'Ticonderoga' class cruisers. The superstructure of the ship is faced with a four-sided aerial array, each side consisting of a 12½ft diameter hexagonal plate covering 4,000 separate active elements. The 90° sectors are fully overlapping, and 360° cover is thus obtained. Comprehensive fault testing and finding is incorporated and the system is robust against battle damage in a way that few other

Soviet air warning radars

Left:
In the 'Moskva' class, the 3D 'Top Sail' is prominent at the masthead but there is also a conventional 'Head Net-C' radar.

Below:
The 'Kiev' class has two 3D radars, both prominent in the photograph: 'Top Sail' before the mast and 'Top Steer' abaft it. Radars lower down in the superstructure are for fire control.

Below:
Radar installation in the *Kirov* is even more complex with the 'Top Pair' back-to-back 3D radar on the foremast and 'Top Steer' on the main. *All Crown Copyright*

Above:
SPS-48C, the standard 3D air warning radar in US carriers, here seen on the after mast of USS *Kitty Hawk*. The aircraft in the foreground are TA-4J Skyhawks, and a F/A-18 Hornet is in the background. *US Navy Official by PHAN R. Eslinger*

Right:
3D Dutch radar fitted in the dome of HrMs *Tromp*.
Rolls-Royce Ltd Photographic Services

radars can be. A smaller but scarcely less effective version, the AN/SPY-1D, is planned for the DDG-51 'Arleigh Burke' class.

In the West at least two other nations are now entering the phased-array field. France has the DRBJ-11, a development by Thomson which is planned for the new nuclear-powered CTOL carrier but will first see service in the 'C70' class anti-aircraft cruisers. In Britain the Plessey company, in what appears to be mainly a private venture, has announced that by the use of gallium arsenide elements, phased arrays can be made much cheaper and lighter, and development may be expected along these lines. The phased array is, indeed, still fairly far down the learning curve and probably presents the most scope for future

Above:
The latest non-rotating phased-array aerials give an uncluttered look to modern ships. In the 'Arleigh Burke' class, a model of which is pictured here, the AN/SPY-1D aerial array is beneath the flat octagonal plate below the bridge. *US Navy Official*

Left:
Bow view of the US nuclear-powered guided missile cruiser *Virginia*, showing the Standard missile launcher. The masthead radar is SPS-48B. *US Navy Official by Newport News Shipbuilding & Drydock Co*

development in shipborne air warning radars. It might in fact be a vehicle for widening the gap between Western and Soviet capabilities; so far, no installation comparable with AEGIS has been seen on the Soviet side, and with their known weakness in miniaturisation they may be late into this field.

Area Defence Missile Systems

Since the middle 1950s the United States has deployed extensively, in its larger surface ships, air defence missiles of considerable range. Terrier, Talos and Tartar were the missiles of earlier generations and all in their day represented a considerable defence against aircraft. All have

Above:
The Standard missile in its most basic installation on a relatively small platform, USS *Rentz* of the 'Oliver Hazard Perry' class. Warning is from the conventional SPS-49 radar on the foremast, and fire control is provided by the STIR (Surveillance & Target Indication Radar) abaft the mainmast together with the Mk 92 system in the dome above the bridge. *US Navy Official by Todd Pacific Shipyard Group*

Left:
Sea Dart, the current area air defence missile of the Royal Navy.
British Aerospace Naval Weapons Division

now been replaced by missiles of the Standard family. These are all semi-active in operation, that is to say their homing heads respond to reflections from the target when illuminated by a radar operated in the launching ship. The shortest-range version is the SM-1, with a quoted intercept range below 20 miles. But there is a two-stage, extended-range version even of this, while the larger and more sophisticated SM-2 too has an ER follow-on that takes its range out to well beyond 50 miles. The missiles use a variety of launchers, all at present of the rotating type, but vertical launch from silos flush with the deck is planned for some classes of ship. The illuminating radars are in the AN/SPG-55-62 series, acquiring the target after indication from warning radar or other means; the larger and more modern the ship the more automated this process is, and the four SPG-62 radars in the AEGIS ships are said to be 'slaved' to the AN/SPY-1 radar.

The US investment in this programme is very great. Over 1,000 Standard missiles were funded in each of the financial years 1984-86, and the retrofit programme for existing classes of ship has proceeded very rapidly. There is no doubt that against aircraft targets at most heights the system has been through a thorough testing programme. Multiple target handling is, particularly in the AEGIS ships, well advanced. Even against attempted saturation attacks by missiles at medium heights, therefore, the AEGIS/Standard system would give a good account of itself and other Standard systems could be expected to perform fairly well. It is admitted that against sea-skimming missiles the prospects are less favourable. American procurement experts have argued in Congressional Hearings that other systems are available to cover that threat. However, the failure of USS *Stark* to cope with a single sea-skimmer — even though human factors may have contributed — could well have reopened that controversy.

Western European navies have also made large investment in area defence missile systems. Britain was early in the field with the now-defunct Sea Slug. This was succeeded in the early 1970s by the much more capable Sea Dart, a fast ramjet missile using semi-active homing on illumination provided by the Type 909 radar. The system proved in the Falklands campaign to be capable of shooting down fast aircraft, and was credited with eight 'kills'. Almost as important was its influence in driving the Argentine aircraft down to wavetop height, where their bombs often failed to arm. But plans for the future development of Sea Dart, beyond minor updating and reliability improvements, are uncertain.

The Royal Navy now holds in Sea Dart a much larger number of indigenously developed and produced area air defence missile systems than any other Western-orientated medium-sized Navy. France still has the Masurca system, incorporating a Mach 3 missile with semi-active homing and a range of 30 miles, in three of its cruisers, but the system is due to phase out in 1995 and apparently no successor is planned. For the rest of its area defence systems, France like Italy and Japan now uses US Standard missiles; in the past these nations too no doubt had indigenous development in mind. Germany, the Netherlands and Spain

Below:

SA-N-3 missile launchers in a Soviet ship of the 'Moskva' class.

never seriously aspired to production of their own systems, and buy American also.

In the Soviet Navy, area air defence missile systems first made their appearance in 1967 with the SA-N-3 'Goblet', fitted in the 'Moskva' class ASW ships. It has been widely fitted since, in cruiser classes and in the 'Kiev' class VSTOL carriers. In its most up-to-date version the missile, fired from a twin launcher and under the control of the 'Head Lights' radar director (probably a command-to-line-of-sight system), is said to have a maximum range of about 35 miles (55km).

The 'Goblet's' line of evolution has been cut across by a quite different missile system, the SA-N-6 'Grumble', which first appeared in the

AREA DEFENCE MISSILES

Designation	*Country of origin*	*Length (ft/m)*	*Weight (lb/kg)*	*Propulsion*	*Guidance*	*Range (nm)*	*Remarks*
Masurca	France	28.2/8.7	4,624/2,098	Two-Stage solid fuel rocket	Semi-active	30	—
SA-N-3 'Goblet'	USSR	20/6.1	1,200/550	Two-stage solid fuel rocket	Radio command	35	—
SA-N-6 'Grumble'	USSR	23/7.01	?	Single-stage rocket	Track-via-missile	50	Vertical launch
SA-N-7	USSR	?	?	Single-stage rocket	Semi-active	15	—
Sea Dart	UK	14.3/4.4	1,212/550	Solid boost Single-ramjet sustainer	Semi-active	30	—
Standard SM-1	USA	14.4/4.47	1,300/625	Solid fuel rocket	Semi-active	20	—
Standard SM-2 MR	USA	15.4/4.7	1,556/706	Solid fuel rocket	Semi-active	25	—
Standard SM-1 ER	USA	26.2/7.98	2,900/1,306	Two-stage solid fuel rocket	Semi-active	40	—
Standard SM-2 ER	USA	26.2/7.98	3,180/1,442	Two-stage solid fuel rocket	Semi-active	80	—

Far left:
SA-N-6 area defence missile silos in a 'Slava' class cruiser.

Left:
A bow view of a 'Sovremennyy' class destroyer. The SA-N-7 single-arm launcher can be seen abaft the twin gun turret. Four of the six controlling 'Front Dome' radars are grouped round the bridge.

'Kirov' class nuclear-powered command cruiser in the late 1970s, and is fitted also in the 'Slava' class cruisers. These missiles are carried in rotating-drum silos with eight missiles in each; the 'Slavas' have eight such silos, the 'Kirovs' 12. Since one missile can be fired from each silo more or less simultaneously, such ships can have up to a dozen area defence missiles in the air at one time. Controlling them would be the 'Top Dome' fire control radar, of which there are two sets in the 'Kirov' and one in the 'Slava' class. Clearly, this should have a multiple target handling capacity, but how flexible it may be in practice is not certain, since the radar is clearly trainable in azimuth and this implies that it might have to deal with targets sequentially. If the missiles possessed their own means of final homing with an onboard active radar, this problem would be eased.

Finally, a successor to the 'Goblet' may have appeared in the SA-N-7 'Gadfly' missile, fitted in the new 'Sovremennyy' class destroyers. Though of shorter range than the latest version of the 'Goblet', it is probably more agile and faster. Working from a single-arm launcher, and guided — probably by semi-active homing — from the 'Front Dome' radar tracker/illuminators, it may sound like a reversion to simpler and less capable systems. But the fact that there are six separate 'Front Dome' installations suggests that multiple target handling is available, and this in turn suggests that the reload and refire rate of the launcher is very rapid.

Point Defence Missile Systems

From their first introduction in the late 1950s, there has been a steadily growing selection of point defence missile systems to choose from in the West. The shorter required range of both missile and tracking sensors, the consequently less stringent requirements for topweight and space, and above all the fact that the missiles were for use against a directly approaching target, led designers and manufacturers to think of the problem as a manageable one in development and production terms by comparison with area defence systems.

The advent of anti-ship missiles, and particularly sea-skimming missiles, as primary targets for point defence systems, produced severe added complications, but did not dampen the enthusiasm of the more robust. Those organisations in the forefront of technology tended to adapt their existing systems to the new threat — which some of them had had in mind from the start — while those with more basic systems preferred to go ahead to a new generation.

The United States occupies nothing like so pre-eminent a place in the point defence missile field as it does in so many other aspects of air defence at sea. Its most widely fitted system is the RIM-7 Sea Sparrow, an adaptation of an air-to-air missile employing semi-active homing. Large numbers have now been procured for the US Navy, examples of the NATO version are in most Western European navies, and some Japanese vessels are also fitted. Missiles are housed in eight-cell box launchers. Several types of tracking and illuminating radar can be used, those most generally in service being incorporated in the US Mk 91 and Dutch WM25 fire control systems.

There are uncertainties surrounding the Sea Sparrow's performance, particularly against small, agile sea-skimming missiles. No direct successor appears to be in prospect, but a very short range anti-air missile is under development in the USA with German and Danish participation. This is the Rolling Airframe Missile (RAM), so called because in flight it is spin-stabilised. It uses a good deal of technology from both Sidewinder and Stinger missiles and can home on both the infra-red and radar emissions of the target missile or aircraft. Once fired it is independent of guidance from the launching platform. This may be a particular advantage against sea-skimmers. Another advantage is the small size and light weight of the missile; it is planned to fit five RAM cells into the Sea Sparrow launch cell, in some installations.

British point defence missile systems are already into their second generation. The first was Sea Cat, originally employing visual-only guidance from a manned director but later acquiring an all-weather capability. Rather short of agility and depending on a good many 'men in the loop' to give it adequate reaction time, Sea Cat is nevertheless a comforting weapon to have on board and is fitted in navies as far apart as Brazil and Indonesia as well as the RN. It was officially credited with eight kills in the Falklands, though some accounts suggest this may be a high figure. There, of course, it was operating against the kind of aircraft targets for which it was designed, rather than missiles for which it was not.

The second generation of British point defence missile certainly was designed with missile targets in mind. This is Sea Wolf, now fitted in over a

D 181
F 209

Left:
The German frigate *Rheinland-Pfalz*, here seen leading line manoeuvres off Portland, has the NATO Sea Sparrow point defence missile system in a box launcher abaft the gun. The Netherlands frigate *Kortenaer*, third in the line, is similarly fitted. *Crown Copyright*

Right:
A British Sea Wolf point defence missile leaves its six-barrelled launcher. *British Aerospace*

Below:
The Sea Wolf is a relatively small missile designed for ease of handling as a 'round of ammunition'. *Crown Copyright by Roger Smart*

dozen ships of the Royal Navy and scheduled for a good many more. It employs command-to-line-of-sight guidance from the Type 910 radar, which gives it great precision against small fast targets; it has hit incoming shells on trials. In its original version this guidance system had limitations against sea-skimmers, and various means — not always satisfactory — were tried in order to improve the situation. The solution now in place is to supplement the main tracker with one of higher frequency to overcome the multipath effects. Such a tracker was available as early as 1976, but since it was a Dutch set it perhaps ran into opposition from British industry.

Other improvements in Sea Wolf are likely to include lightweight launchers to replace the original rather cumbersome and heavy six-cell hand-reloaded model, or — in the RN's Type 23 frigates — a vertical-launch system from 32 deck silos. The system was designed from the outset to have very quick reaction and its response requires very little human intervention. It is claimed to be, and almost certainly is, the best anti-missile missile system available in the West.

The current French point defence missile system is the Crotale, an adaptation of a land-based missile. Its guidance is by beam-riding and it is detonated by an infra-red fuze. The original Crotale had limitations against very low and crossing targets, and the update due into service in the second half of the 1980s will seek to give it the ability to handle Mach 2 targets down to 15ft. An infra-red tracker will be the main new feature. Crotale is fitted in many major units of the French Navy including the aircraft carriers; it is interesting, and rather mystifying, that British carriers of the 'Invincible' class carry Sea Dart but not Sea Wolf.

Several new systems are projected in France. The *Système Surface-Air à très courte portée* (SATCP), incorporating infra-red and laser homing techniques and using a tungsten-ball warhead, is expected in service in the late 1980s, while a few years further on there is the possibility of a whole family of systems called generically SAMAT (*Surface-Air Matra*). These are claimed to feature very agile missiles employing infra-red seekers in SAMAT 1 and an active-radar homer in SAMAT 2.

Italy's most generally fitted PDMS is the

POINT DEFENCE MISSILES

Designation	*Country of origin*	*Length (ft/m)*	*Weight (lb/kg)*	*Propulsion*	*Guidance*	*Range (nm)*	*Remarks*
Crotale	France	9.5/2.89	185/83.9	Solid fuel rocket	Command/infra-red	10	—
SATCP	France	5.9/1.8	37.5/17.0	Solid fuel rocket	Command/infra-red	3	Under development
SAMAT	France	?	?	Rocket	Infra-red (1) Active radar (2)	5+	Under development. Two guidance versions
Albatros/ Aspide	Italy	12.1/3.7	485/204	Single stage rocket	Semi-active	8	Compatible with Sea Sparrow. Exported to 15 countries
Barak	Israel	8.2/2.5	192/57	Rocket	Semi-active	8	Designed for fitting in small craft
SA-N-1 'Goa'	USSR	22/6.7	882/400	Two-stage rocket	Radio command	9	In service 1961
SA-N-4 'Gecko'	USSR	10.5/3.2	419/190	?	Radio command	8	Retractable pop-up launcher
SA-N-5 'Grail'	USSR	4.75/1.45	32/14.5	Rocket	Manual aim, irfra-red homing	5	Light forces, auxiliaries and landing ships
SA-N-8 and 9	USSR	?	?	?	?	10?	New construction
RIM-7H Sea Sparrow	USA	12/3.7	500/227	Solid fuel rocket	Semi-active	12	Fitted also in many W. European navies
RAM	USA	?	?	Solid fuel rocket	Passive RF/ infra-red seeker	3	Under development. 5in diameter. W. German participation
Sea Cat	UK	4.9/1.47	138/68	Two-stage rocket	Optical	3.2	Radar or TV guidance in later versions
Sea Wolf	UK	6.5/1.9	176/82	Solid fuel rocket	Radio command to radar-line-of-sight	5	—

Albatros Aspide, which though produced entirely in Italy incorporates many features of the Sea Sparrow and shares its main characteristics. It is controlled by the NA-30A fire control system. Albatros, or further adaptations of it, is having much success in export markets including that of Japan.

Israel has lately entered the PDMS field with the Barak system. Its salient characteristic is that it is very light and therefore available for fitting into vessels of as little as 350 tons. It employs semi-active homing from an illuminating radar based on an airborne set and employing modern ECM techniques. Microprocessing is extensively employed in the fire control system and this, with the claimed agility of the missile, is expected to cancel out many of the known limitations of the semi-active principle against very low targets.

If the West has been busy in the field of PDMS, the Soviet Union has certainly not been idle. There appear to have been three generations, all widely fitted in their day and all still operational in some classes. The SA-N-1 'Goa' was introduced in 1961 and is found in the cruiser and destroyer classes of that time. It employs command guidance from the 'Peel Group' radar director. It is probaby effective against aircraft targets down to about 500ft, but has severe limitations against missile targets.

The second generation is a more agile and

sophisticated affair. This is the SA-N-4 'Gecko', fired from a retractable launcher linked to a 20-missile magazine. Command guidance is again employed, this time from the 'Pop Group' director which gives a capability against much lower targets — down to 30ft, it is said. The generation of the 1970s and early 1980s, from the first of the 'Kirov' class battlecruisers to 'Nanuchka' class corvettes, is fitted with this system.

Finally, the SA-N-9, fitted in the later 'Kirov' and 'Kiev' classes and the 'Udaloy' class destroyers, is a vertical-launch system about which little is yet known but which can be expected to have all the capability of the 'Gecko' plus multiple target handling, though it is unlikely to be a 'fire-and-forget' system like the RAM or SAMAT.

Gun Systems

Mark Twain once said that reports of his death had been exaggerated. If naval guns had been able to

Below left:
The vertical launch version of Sea Wolf.
British Aerospace

Above:
In the Soviet cruiser *Slava*, the SA-N-4 point defence missile system and associated 'Pop Group' directors are well aft, sited port and starboard at the break of the quarterdeck.

Centre right:
SA-N-4 is fitted in craft as small as the 770-ton 'Nanuchka' class.

Right:
SA-N-9 vertical launch point defence missile system silos forward of the gun mountings in the Soviet anti-submarine destroyer *Vitse Admiral Kulakov*.
Crown Copyright

Above:
A 130mm dual purpose twin gun in a 'Sovremennyy' class destroyer.

Below:
The Creusot-Loire Compact 100mm gun.
Creusot-Loire Industrie

use words, no doubt they would have said much the same thing. Gunless ships and indeed gunless navies were quite seriously and generally canvassed in the dawn of the missile age. However, two or three decades on, there is if anything a tendency to fit more gun systems in new construction, principally for air defence.

The reasons are fourfold. First, the threat of saturation attack is regarded as a very real one, coming not from the Soviet Navy alone but from smaller forces with heavy anti-surface-ship missile armament. This threat demands many channels of fire in the defender. Secondly, the concept of layered defence gives a clear role for the gun at a point in the action where the PDMS has either failed or proved inadequate. Thirdly, gun design has recovered from a plateau in the learning curve and continued to advance. Finally, gun systems are generally lighter and cheaper than current PDMS.

Naval guns now fall into four broad categories. First are the heavy guns, down to 6in or so, rare indeed now and even more rarely used for air defence. There are schemes for laser guidance of shells from such guns in the US Navy but the opportunities for such use are unlikely to occur frequently. They will not be discussed further. Secondly comes the category of guns between 4in and 5in (100-130mm). These generally have a multiple role, shore bombardment and surface action being the considerations apart from air defence. Thirdly come the guns around 3in (57-76mm), which with their more rapid rates of fire and lighter shell are predominantly for air defence but have some surface action role. Finally, the guns below this calibre are of short range, have typically very high rates of fire and are provided primarily for air defence including, in their most modern versions, anti-missile defence.

No less than five Western nations produce modern guns in the 100-130mm category: France, Italy, Sweden, the UK and the USA. The three most important variables in the performance characteristics of such guns are rate of fire, weight of shell and maximum range. It is not surprising that the first of these varies inversely, and the other two more or less directly, with the calibre. Thus the French Creusot-Loire Compact of 100mm has a rate of 90 rounds per minute but only throws a 30lb shell, while the 127mm Italian Oto Melara has a rate of fire of 40 rounds per minute but throws a 70lb shell, and has a range half as much again as the Compact.

When choices have to be made, further parameters have to be considered. Quite apart from trade-offs in the fields of shore bombardment and surface action, it is necessary to think about topweight (small light mountings may be imperative for corvette-sized craft), crew (the unmanned mode of the UK Vickers mounting may be

NAVAL GUNS, 100-130mm

Calibre/ length	Country of origin	Elevation	Rate of fire (rpm/barrel)	Weight of shell (lb/kg)	Maximum surface range (km)	Remarks
130/70	USSR (1980)	85°	35	?	28	—
127/54	USA (1975)	80°	20	70.5/32	24	Mk 42 Mod 9, a later variant, has higher rate of fire
127/54	Italy (1968)	85°	40	70.5/32	24	—
120/46	Sweden (1967)	80°	80	46.3/21	18	—
114/55	UK (1971)	55°	25	46.3/21	22	—
100/55	France	80°	90	29.8/13.5	17	Creusot-Loire Compact
100/60	USSR	85°	80	?	18	—

NAVAL GUNS, 76MM AND BELOW

Designation	Calibre (mm)	Country of origin	Barrels per mounting	Rate of fire (rpm/mounting)	Remarks
Oto Melara Compact	76	Italy	1	85	Effective range in AA fire 3nm. Widely fitted in Western navies, including USN under licence
Dual purpose	76	USSR	1 or 2	90-120	Effective range in AA fire 6nm. 'Bass Tilt'/'Owl Screech' fire control
Model 1951	57	France	2	120	—
Dual purpose	57	USSR	2	240	Effective range in AA fire 3nm. 'Muff Cob'/'Bass Tilt' fire control
Bofors SAK57	57	Sweden	1	220	Effective range in AA fire 5nm
Breda/Bofors	40	Italy	2	600	Effective range in AA fire 2½nm. Fire control by Dardo system
Bofors Trinity	40	Sweden	1	330	Programmed spread of aiming points with high lethality shell designed to ensure high kill probability
Oerlikon	35	Switzerland	2	1,100	High muzzle velocity and slew rates
Rarden LS-30B	30	UK	1	650	Type 23 frigates. Oerlikon/BMARC gun
Goalkeeper	30	Netherlands	7	4,200	'Kortenaer' class; also ordered by UK. Track-while-scan radar fire control. GAU-8 Gatling gun
Gatling	30	USSR	6	3,000	'Bass Tilt' fire control
Breda/Mauser	30	Italy	2	1,600	Designed for fitting in light craft
Oerlikon	30	Switzerland	2	1,300	Enclosed aimer's cabin incorporated in mounting
Oerlikon	25	Switzerland	4	3,400	Designed as anti-missile gun. Incorporates inclined traverse bearing concept
Light AA	25	USSR	2	?	In older construction
Vulcan/Phalanx	20	USA	6	3,000	Radar-controlled by matching line of sight/projectile stream. Under continuous development. Depleted uranium/tungsten ball ammunition
Oerlikon	20	Switzerland	1	800	Very light weight for fitting in small craft

attractive), the compatibility of fire control systems, and deck magazine arrangements, particularly the availability of ready-use ammunition. If, finally, the choice falls on the system which is most attractively packaged or offers most opportunity for indigenous participation in its production, that may be no great surprise.

The guns in the 57-76mm category are found in an even wider variety of navies. Indeed, a roll-call of the navies fitted with the most widespread, the Oto Melara 76mm, runs into nearly a couple of score. The reasons for the popularity of such weapons are many. First, the mountings are relatively light and therefore can be fitted in quite small craft, of the kind with which many countries first equip small navies. Secondly, they are simple to maintain and operate. Thirdly, they can be mated with a variety of fire control systems through easily managed servo units. And finally, of course, supplies of spares and ammunition are widespread.

In fact, the rates of fire of these weapons are not always much in excess of the quickest-firing guns in the heavier category, but their handiness and lightness is enough to preserve their popularity. They do also have some surface-action role.

The category of small-calibre guns, less than 50mm, splits really into two. There are the guns which are carried simply to create one further channel of fire, often quite crudely controlled by visual aiming and manual elevation and training; the simpler versions of the Oerlikon family are of this type, and in good hands may be capable of exerting a deterrent influence against light threats including intrusive helicopters or even fixed-wing aircraft. But there is a much more sophisticated

Above:
Guns of both 76mm and 40mm calibre (Oto-Melara and Breda respectively) in a Tunisian patrol boat. *Thomson SDC*

Right:
Typical of modern multi-barrelled close-in gun systems: the Dutch Goalkeeper.
Hollandse Signaal Apparaten

range of weapons now becoming available, which have an anti-missile role, at least in intention.

Typically, these are largely self-contained systems with their own built-in directors and fire control. It may be possible to control them from other stations in the ship, but that would be a secondary mode. They are also designed to need minimal human intervention; in the primary mode, indeed, most of them would go into action after the indication of a missile target without any further order. For example, the Dutch Goalkeeper system, it is said, 'searches the sky continuously, ensures timely detection and classification of any potential threat, selects and designates the highest priority target, predicts accurate gun-aiming point, opens fire, spots and corrects, performs kill assessment and remains ready for immediate engagement of the next target, all automatically'. That is quite a claim for a system with an all-up weight of 6,000kg (5.9 tons). But it is not unusual. The US Vulcan/Phalanx and the Swedish (Bofors) Trinity systems, using very different technical solutions, both offer similar comfort as last-ditch anti-missile systems.

As with all such very close range weapons, one snag is that a missile, even though hit, may fly through and still impact on the ship. This leads to a search for ever more accuracy in aiming and lethality of ammunition. Many of the advances originally made in anti-tank warfare have been translated into the ammunition now fitted in these systems.

As might be expected, the Soviet Navy has a full range of anti-aircraft guns in all the categories mentioned above. Moreover, it fits them in great quantity, Soviet ships being in general much more heavily gunned than their Western counterparts. Systems are available in 130mm, 100mm, 76mm, 57mm and 30mm calibres; in some cases the fire control may not be of the most modern, but for example the 30mm gatling mount, with six barrels and controlled by the 'Drum Tilt' radar director, can put into the air 3,000 rounds per minute — not far short of the 4,200 rounds per minute claimed for the Goalkeeper.

Electronic Warfare

Manufacturers — and governments even more so — are reluctant to discuss details of electronic warfare equipment. However, enough information has been published to give a summary of the principles on which modern ship-fitted EW equipment works. There is ever closer integration between the ESM, or detection, side of electronic warfare and the ECM, or jamming, side, and the two will be considered as part of one EW system, even though ESM information is of course used more widely in the general information picture on board and in the force.

Typically, a ship-based EW system such as the American SLQ-32(V)3, the French ARBR 17/ARBB 33 or the British RAPIDS/SCIMITAR consists of a radar search receiver covering the whole 1-18GHz band (B to J bands), employing either a multi-port or rotating dish aerial. In order to measure the frequency of received signals, an instantaneous frequency measuring device, using a swept or stepped local oscillator, is employed. The bearing, amplitude and pulse width of the received signal are also measured. These parameters can then be compared with an electronically stored 'library' of characteristics of known threat radars.

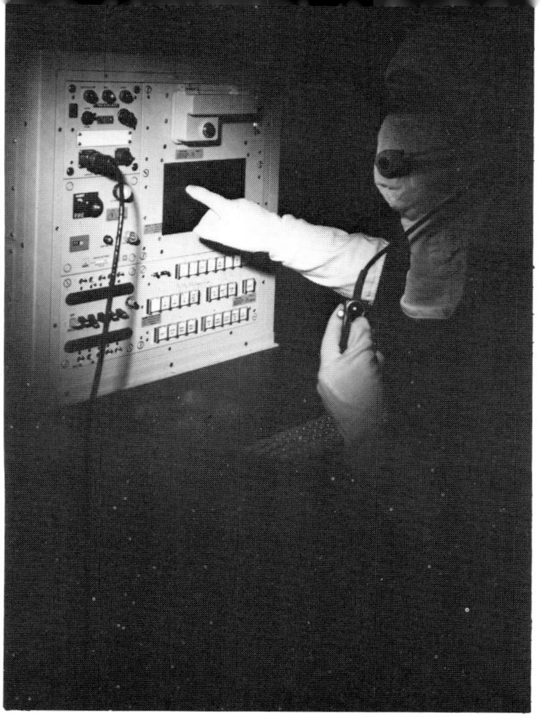

This information can be used in several ways. First, it can be passed to the rest of the force so that they are warned and may be able to fix the transmitter by cross-bearings. Secondly, it gives the command some idea of the immediacy of threat; clearly, a signal with the known characteristics of a homing sea-skimmer is more urgent than a shadowing aircraft's search radar. Finally, it provides the basis for Electronic Countermeasures (ECM).

The ECM (jamming) side of the EW equipment responds either to command instructions or, in the case of urgent threats and in the most modern equipments, directly to inputs from the ESM side. The 'library' characteristics are passed to the jammer which searches for them both directionally

Below left:
Soviet version of the multi-barrelled gun: the widely-fitted 30mm Gatling.

Below right:
Two 'last-ditch' systems: the Oerlikon 25mm single mounting with, in the background, a chaff launcher. *Oerlikon Military Products*

Above right:
A Guardian electronic warfare console.
Thorn EMI Electronics Ltd

and in frequency. Once a target matching the designated characteristics has been acquired, the jammer locks on and begins a pre-programmed response planned to cause the maximum confusion to that particular target radar. The techniques involved go by a series of names such as Range Gate Pull Off and Inverted Scan Rate Modulation; it is easiest to think of them as echoes with false ranges or bearings, causing an incoming missile to follow them rather than the response from one's own ship.

The Soviet Navy is known to pay much attention to electronic warfare in all its forms, but it is also very circumspect in its use of jammers — and the West is equally cagey about saying what it knows of Soviet equipment. Thus it can be seen that every Soviet surface ship of any size is fitted with both ESM and ECM gear, and the NATO codenames — 'Bell Clout', 'Watch Dog', 'Rum Tub' and so on — are known; but published information on the uses and the efficiency of the equipment is not available.

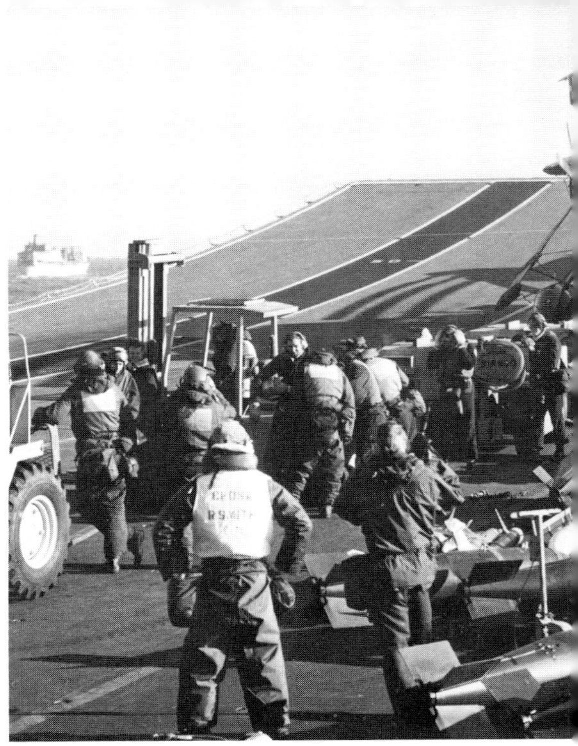

Decoys

Chaff decoys against radar were widely used in World War 2. The basic device has not changed: it is a mass of strips of foil cut to length to maximise their reflecting properties against certain frequencies of radar, released into the air at points where they will give maximum protection.

It has been well said that there are three modes of operation for decoy systems: confusion, to cause doubt in the enemy's mind about the size and shape of a force; distraction, to produce false targets for him at the shadowing and acquisition stage; and seduction, to lure missiles, once fired, away from true targets on to false ones.

The confusion mode requires a dispensing system of relatively long range — typically several miles. The most common form of dispenser is a shell fired from a gun. Many navies have such ammunition, but every shell of this sort means one less of other sorts in the magazine, so stocks will be limited.

The distraction mode also requires relatively distant dispensing, perhaps a couple of miles from the ship. Rocket launchers are commonly used for this. A sufficient quantity of chaff must be released to simulate a ship of about the same size as the vessel to be protected; generally this will be one's own ship, in which case the pre-planning is not too difficult.

The seduction mode is more risky. In the most usual operation of this mode, the ship envelops itself in chaff and then steams as quickly as it can upwind and out of the cloud, hoping that the missile will continue to home on the chaff and not return to its ship target. Remembering the time available to counter a sea-skimming missile approaching at Mach 1, it can be seen that the miss-distance is likely to be very small and those few seconds will be nail-biting.

Equipments available in the West include the American SRBOC system, the British Plessey BBC and the French Dagaie/Sagaie system. Some of these systems now incorporate flare launchers, designed to seduce missiles which have infra-red devices either as primary or supplementary homers.

Logistic Considerations

The vast majority of air defence systems at sea depend on onboard stocks of one kind or another. Ammunition for hard-kill systems, whether missile or gun, is the most obvious example. Even in the most modern United States ships, only 68 Standard missiles are carried. The Soviet Navy is thought to do rather better with 96 SA-N-6 in the 'Kirov' class. British and French area defence missile ships carry a maximum of 40 missiles per ship — in some cases less. The story is much the same, and the actual numbers scarcely greater, for point defence systems. Vertical launch, when available, will not make much difference. Gun ammunition is unlikely to run to much more than 400 rounds per gun in the 120-130mm bracket; for smaller calibres the figure is commensurately higher, but so is the rate of fire.

Decoys also tend to eat up stocks at a prodigious rate. The firing of chaff rockets, particularly in the distraction or seduction modes, has to be an instant-reaction drill at the first indication of

Left:
This well known photograph of HMS *Invincible's* flightdeck during the Falklands conflict illustrates the scale of the logistic problem in any major seagoing operation. The Royal Fleet Auxiliary (left background) was never so sorely needed, and back-up from the Merchant Navy was essential.
Crown Copyright

Below:
A schematic diagram of a Weapon Control & Action Information system in a small ship.
Ferranti Computer Systems

immediate missile threat; if it is not, it is likely to be ineffective. Moreover, since it demonstrates no hostile intent, the rules of engagement covering its use are likely to be liberal. In consequence, in an area of air threat, the expenditure from launchers at any level of hostilities is likely to be high.

Even electronic devices, whether warning radar, fire control, ESM or ECM, demand spare parts from time to time. If these are not immediately available on board, life can become extremely uncomfortable for the command; nothing gives such a naked feeling as a vital piece of electronic equipment that is 'down'.

There are several implications. First, a high premium is put on minimising false alarm rates. In conflict, it simply will not be good enough to blaze away at every contact that is picked up and might, or might not, be an approaching threat. Secondly, every effort must be made to improve ship design so that on-board stocks can be increased. Crew comfort has to be balanced carefully against the potential for survival. Finally, afloat logistic support — 'the fleet train' — assumes outstanding importance in an area of air threat; fighting units must be able to take from auxiliaries not only fuel

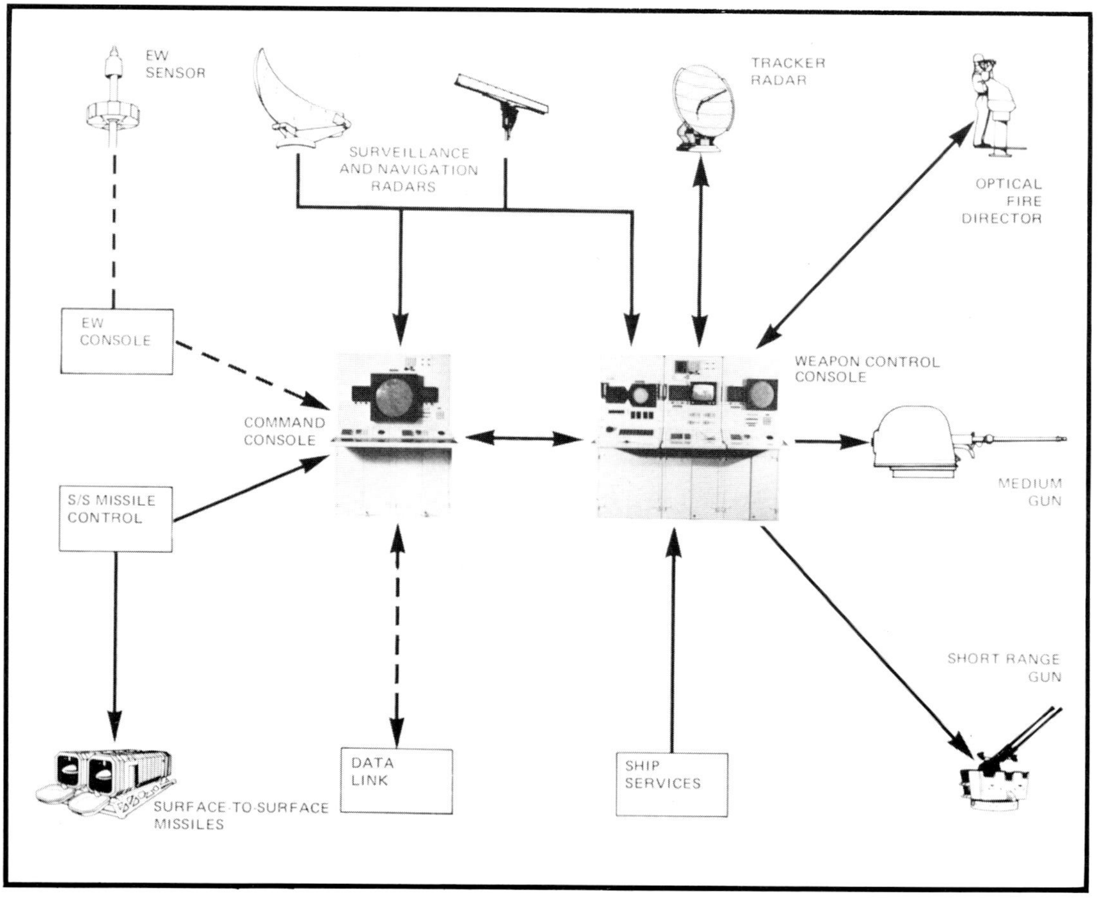

but all kinds of stores and ammunition, under way and in a wide range of weather and sea conditions. The USA and the United Kingdom have the most advanced capabilities in this field, most of the other navies of the West are competent but not well supplied with logistic ships, and the Soviet Union still has much to learn.

Putting it Together

Air defence warning, weaponry, countermeasures and logistics in a major force add up to an enormous investment which in turn is intended to promote the survival of an even greater investment. It is manifestly necessary that these assets should be used in the optimum way when a threat develops. Each threat must be engaged by the most suitable system, or combination of systems, from the most appropriate unit.

It is no surprise, therefore, that much thought and money has been put into the development of command, control, communication and information (C^3I) systems. The basis of all such systems is now the digital computer into which raw information is fed and where it is processed, compared and prepared for presentation on electronic displays to the command and to other users. Increasingly, multiple computers linked by data highways are employed; this reduces dependence on a single processor with the attendant chances of catastrophic loss of data in the event of breakdown.

The number of Western nations producing such C^3I systems is quite large. The American Naval Tactical Data System (NTDS), the British CAAIS/CACS series, the French SENIT and TAVITAC, the Dutch SEWACO, and the Italian SADOC, are modern examples; some earlier designs are still in service. The proliferation of these systems may present compatibility problems demanding interfacing equipment, though digital techniques do simplify matters and it is with the earlier equipment that most difficulty may be found.

The Soviet Navy must, of course, have such systems as well; indications are that they are more decentralised than in Western navies. This is surprising in view of the known Soviet tendency to tight control. It would be extremely interesting to walk round the *Kirov's* operations room at the height of a major exercise. Most reports from sea suggest that however it is done, reaction to the approach of Western units is generally rapid and co-ordinated; but this contrasts with occasional instances of appalling doziness and sloppy drill.

Like everything else, it probably boils down in the end to who is on watch; and their actions will depend not only on what sort of people they are, but how they have been trained and even what they've had for breakfast. The best gear and weaponry in the world will not be effective if it is not properly organised, maintained and served.

Left:
The command positions in a Type 22 frigate.
Ferranti Computer Systems

Right:
Close-up of a command console.
Ferranti Computer Systems

Below:
Training at all levels is a prerequisite for effective air defence.
Crown Copyright

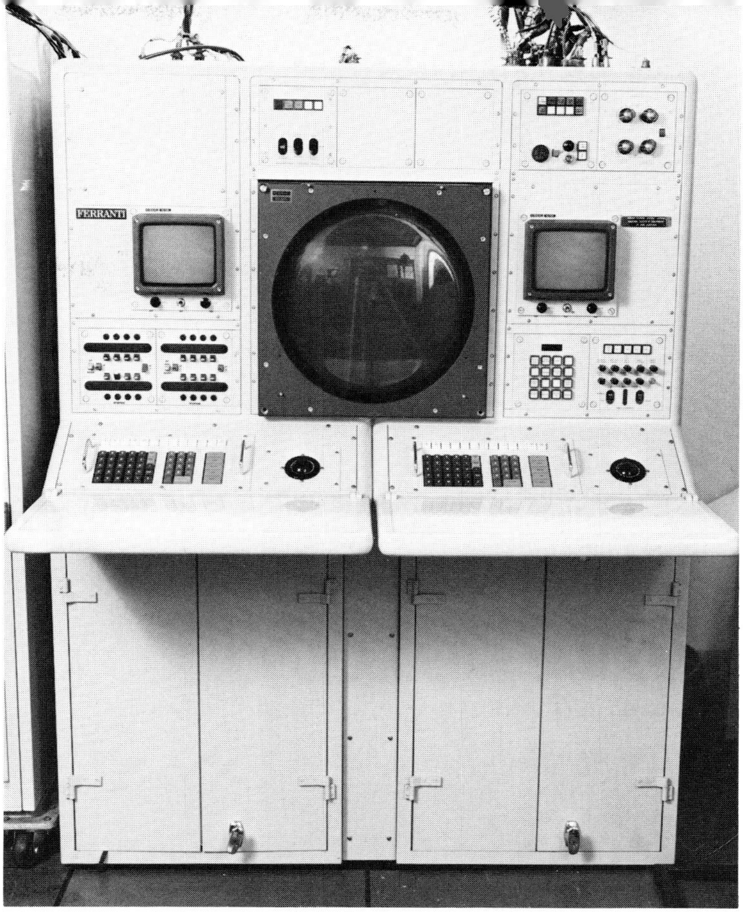

CHAPTER 4

Two Air Defence Battles

This chapter will discuss how the principles and the weaponry of air defence might be applied in two sea battles: the first a 'classic' NATO-versus-Soviet encounter, of the kind we all hope deterrence will serve permanently to prevent; and the second a medium-power operation against moderate opposition, which will consider mainly the air defence side of the Falklands experience as it has emerged from books and some discussion on the subject.

Some general considerations apply to both. First, the aim of an operation must be kept always in mind. If there is no strategic necessity for a maritime force or unit to be in an area of high air threat, then it should not be there. 'Strategic necessity' can cover a wide range of requirements; for example, a force can be sent into a high-threat area deliberately to draw fire from hard-pressed forces elsewhere. But in general sea forces exist, and are operated, to secure the use of the sea in order that the totality of a strategy shall be successful. Their exposure to air threat will grow from this function.

Secondly, the air defence battle can never be considered in isolation. In the pages that follow it will naturally dominate the discussion, but it must be remembered — as the commanders at sea and in shore headquarters must always remember — that the threat of underwater weapons is generally present in modern higher-level operations. This will not only affect the disposition of the surface and submarine units, but will complicate air operations because of the need to conduct airborne anti-submarine warfare.

Finally, it will not be possible for reasons of space to discuss in any detail the air defence aspects of moves leading up to the outbreak of generalised hostilities — before the transition, in my own jargon, from low intensity to higher level operations. The former will be conducted under strict rules of engagement, with definitions and assessment of hostile intent uppermost in the minds of commanders, great attention paid to surveillance, much circumspection in the operation and emission policies of the main forces, and probably deceptive measures in the disposition and profile of subsidiary forces. Individual units will be keyed up against the possibility of an initial onslaught by submarine or air or both, possibly for day or weeks on end: a tense, anxious time.

But it is assumed in what follows that that time is about to come to an end, unambiguously, and that Rules of Engagement are relaxed to those necessary to deploy air defence resources in the most effective way. So, as the Great Duke said when Sam Small had picked up his musket, 'Let battle commence'.

The NATO Encounter

Because it brings in all the salient factors of air defence in general, in a way scarcely any other scenario does, the battle to be considered is one in the Norwegian Sea, in support of what has come to be called The (with a capital 'T') Maritime Strategy declared by United States Secretary of the Navy Lehman and Admiral James Watkins in January 1986. In turn The Maritime Strategy is based on the principles of deterrence, flexible response and (particularly) forward defence, which are the tenets of NATO strategy.

The avowed purpose of the Strategy is to pin back the Soviet Northern Fleet in its bastion and allow it no respite or refuge there. In this way, it is argued, the Soviet maritime forces will be constrained to pay great attention to the preservation of their ballistic missile submarines, which they need to keep as a second-strike force and even possibly as an end-of-conflict bargaining chip. They will also be concerned to protect the Kola base itself. Thus they will have few forces to spare to attack NATO sea lines of communication, conduct amphibious operations or in other ways support the land battle in Europe. Conversely, NATO will be able to make use of the sea in a way which alone can bring a land battle in Europe to a successful conclusion.

The Strategy, however, is by no means risk-free. It embodies two concepts which historically have been the prerogative of the stronger naval power:

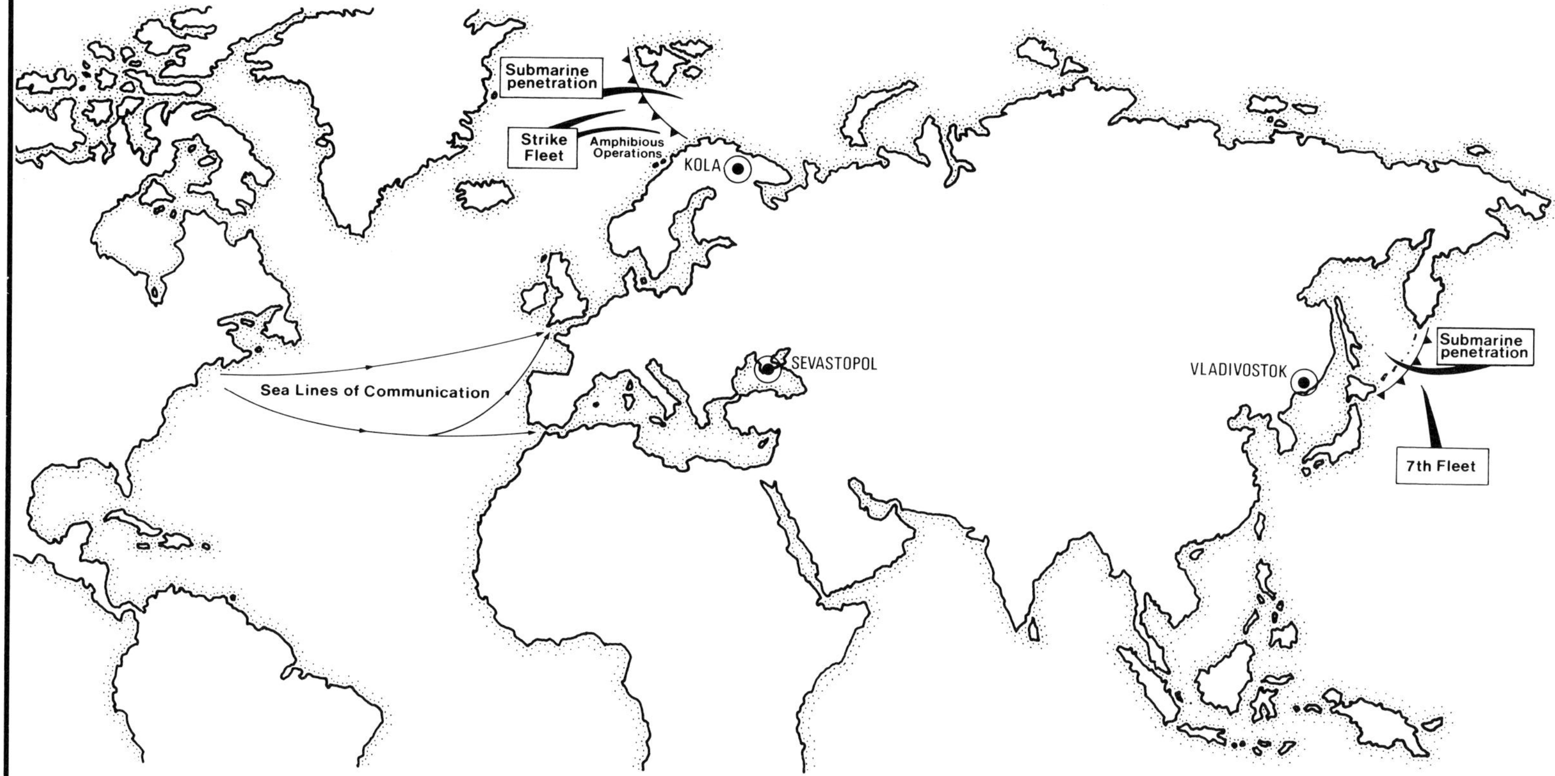

THE MARITIME STRATEGY

blockade of the enemy fleet, and seeking a decisive battle. Such a policy was adopted by British strategists in the Napoleonic Wars, where both geography and the technology of the time were in its favour; but even the Grand Fleet in World War 1, although it certainly regarded itself as ready and willing to fight a decisive battle, did not attempt blockade because the submarine and the mine proved themselves to be such effective counters. The United States Navy would, to be sure, not need to encounter too many mines in the open blockade the Strategy envisages; but it would have to cope with submarines; and, even more, would have to absorb the severe air threat that has been described in this book.

In order to fight this battle successfully, therefore, very strong forces will be needed. All indications are that the Americans would want to put at least three CVBGs into the Norwegian Sea with support from a British ASW Carrier Group and from land-based aircraft operating from Iceland, Norway and the United Kingdom. Amphibious Groups might be operating in the area also, depending on the situation ashore and how it had developed; had matters gone well, the initial landings in support of Norway might have taken place before hostilities started. But even then, resupply by sea of such forces would be going on.

This then is the scenario at the start of serious

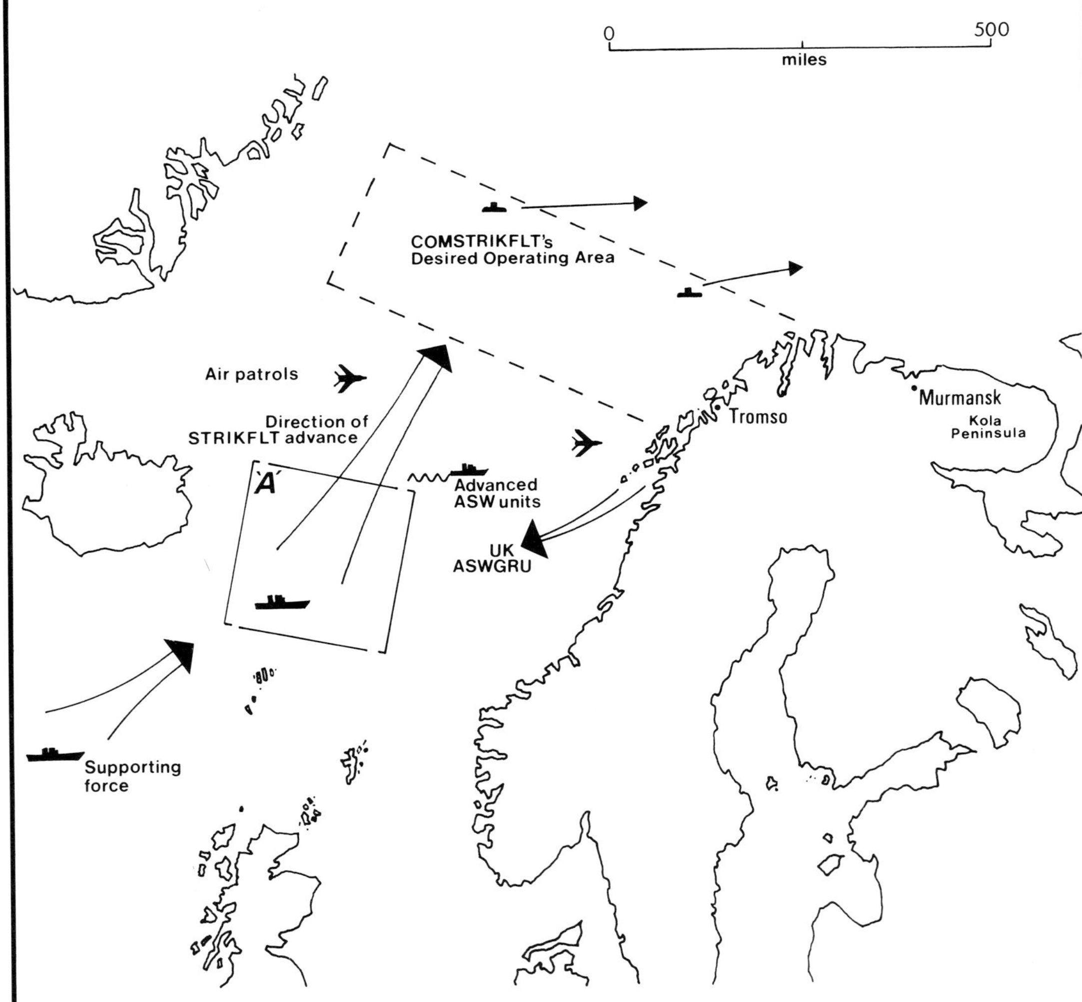

THE NORWEGIAN SEA SCENARIO: a possible NATO plan

Components of COMSTRIKFLT's force

Left:
F-14A Tomcat fighter aircraft.

Below:
AEGIS missile cruisers.

Bottom:
Strike and defensive power in the carriers.

Left:
Highly sophisticated command, control, communication and information systems.

Below:
A general-purpose destroyer.
All US Navy Official

hostilities. The USN deploys, just north of the Greenland/Iceland/Faeroes Gap, two CVBGs, consisting each of a strike carrier, two AEGIS ships and four other cruiser/destroyer types. A third CVBG is approaching the gaps from the south. The UK ASW group, consisting of an 'Invincible' class carrier, two Type 42 destroyers and four Type 22 frigates, has just finished covering a successful unopposed (though closely observed) landing in the Troms area and is withdrawing southward to replenish. Further forward still are the units conducting more independent anti-submarine operations: submarines themselves, both American and British, and towed-array surface units, somewhat lonely and exposed. Intensive air patrols, both ASW and anti-surface vessel, are being conducted by shore-based aircraft.

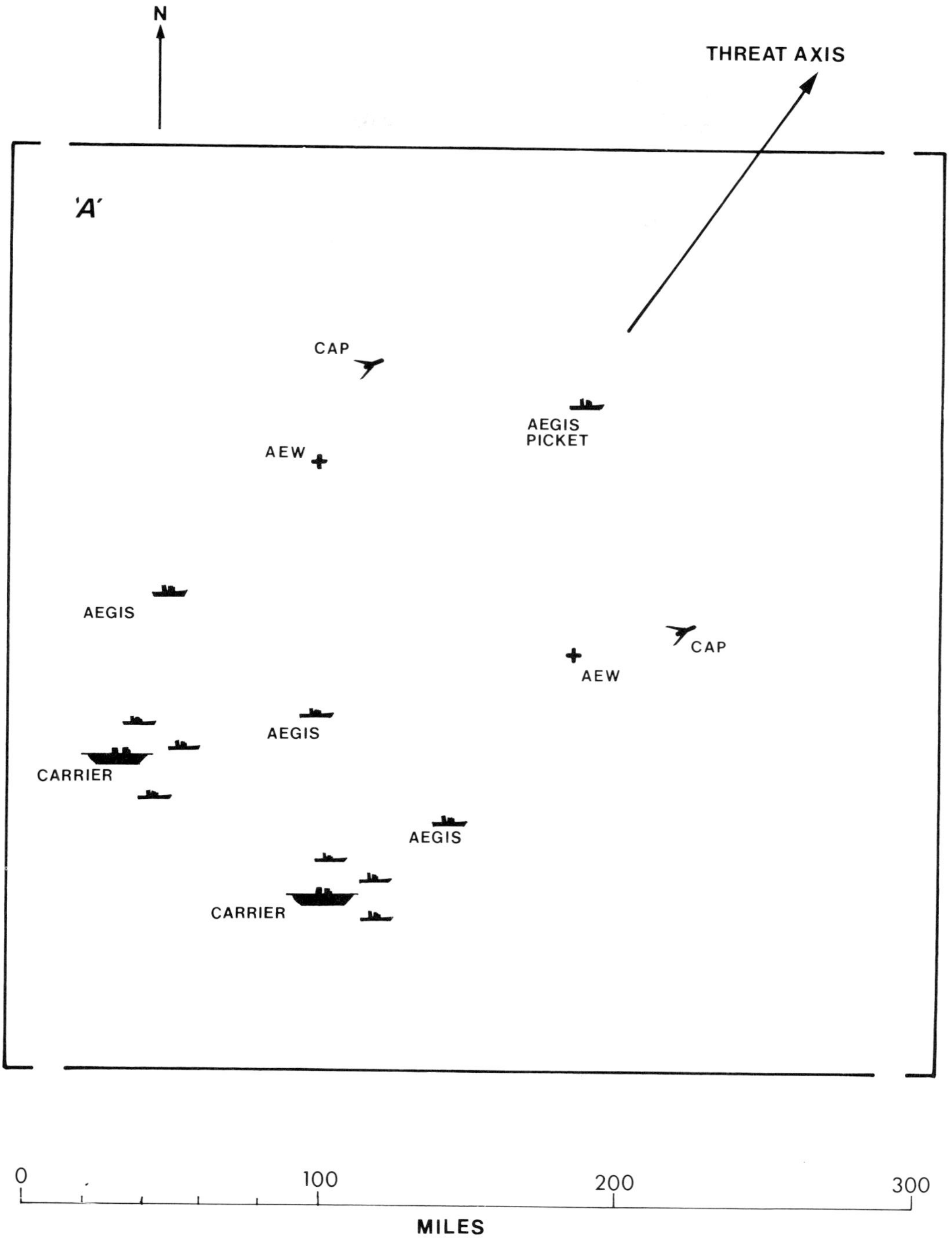

THE NORWEGIAN SEA SCENARIO: Possible Strike Fleet Air Defence Disposition

(Detail of area 'A' in diagram on page 86) Forces specifically for ASW not shown

It can be taken that the Commander of the US Strike Fleet has overall tactical command in the area under the Commander-in-Chief Eastern Atlantic, although that is something of a simplification: the actual command structure is complex and involves numerous authorities and procedures for the assumption of control of a particular unit by the most appropriate commander at any given time.

COMSTRIKFLT's immediate objective is to establish his CVBGs between the latitudes of the Arctic Circle and North Cape. From here he can support the land campaign in northern Norway which he expects shortly to develop; can threaten the Soviet bases in the Kola and, indeed, Arkhangelsk areas; and can support his forward-deployed anti-submarine forces, both sub-surface, surface and air.

But he does not expect to get there without a fight. Soviet activity is reported to be intense and the intelligence reading is that general hostilities are imminent. There is, moreover, firm evidence from Soviet activity in the vicinity of the CVBGs that their position is known and their composition and disposition accurately reported.

In consequence COMSTRIKFLT assesses that nothing is to be lost by operating his aircraft in the optimum air defence mode, and by a relatively free emission policy for communications, except for HF. However, he continues to restrict the use of air warning radar by ships, preferring to rely on his AEW aircraft for the time being. The exception is a single AEGIS-fitted picket detached 180 miles to the north of the main force; this, it is hoped, will make enough noise to confuse the opposition, particularly as it has orders to shift station and use various transmission modes. COMSTRIKFLT wishes he could spare two ships for this job. However, he has his carriers about 80 miles apart, and assesses that for their defence he needs three AEGIS ships across their front at a distance of some 50 miles. The other destroyers and frigates are closer in, with a dual air defence and ASW role.

A threat axis is established on the assumption that Soviet strikes, when they develop, will seek to avoid flying through the cover of the northern Norwegian shore-based NADGE radar chain. AEW patrols, one from each US carrier, are disposed either side of this axis and about 120 miles up-threat from their respective parents. Each has airborne a Combat Air Patrol of four aircraft, F-14s and F-18s alternating in this role, under AEW control. These CAPs are stationed even further out, about 150 miles from their carriers and again up-threat. A 2hr 20min flying cycle, with fast transit to and from CAP stations, is giving the flying operations staffs some concern; they wonder how it will stand up to action conditions if it comes to that.

Weather conditions contain some good and some bad news. The good news is that cloud cover is heavy, impeding visual observation and positive identification by aircraft or satellite. The bad news is that this goes with a west-southwesterly wind, necessitating flying courses in that direction and reducing the speed of advance of the carriers. Rain, sea clutter and cloud are not going to help detection of small low targets either.

Meanwhile, the British force is doing its best to work in under the umbrella of the US fighters. An into-wind replenishment course helps. The very small Sea Harrier force is being kept at deck alert, but one AEW helicopter is airborne towards the Norwegian coast in order to detect any Soviet aircraft attempting an overland transit to outflank the NADGE chain to the south.

The slow movement of NATO forces into a more advantageous position goes on throughout most of an uneasy day. In view of the steadily worsening political intelligence, COMSTRIKFLT presses for further relaxation in the Rules of

Below:
An EA-6B Prowler aircraft from Electronic Warfare Squadron 137. *US Navy Official*

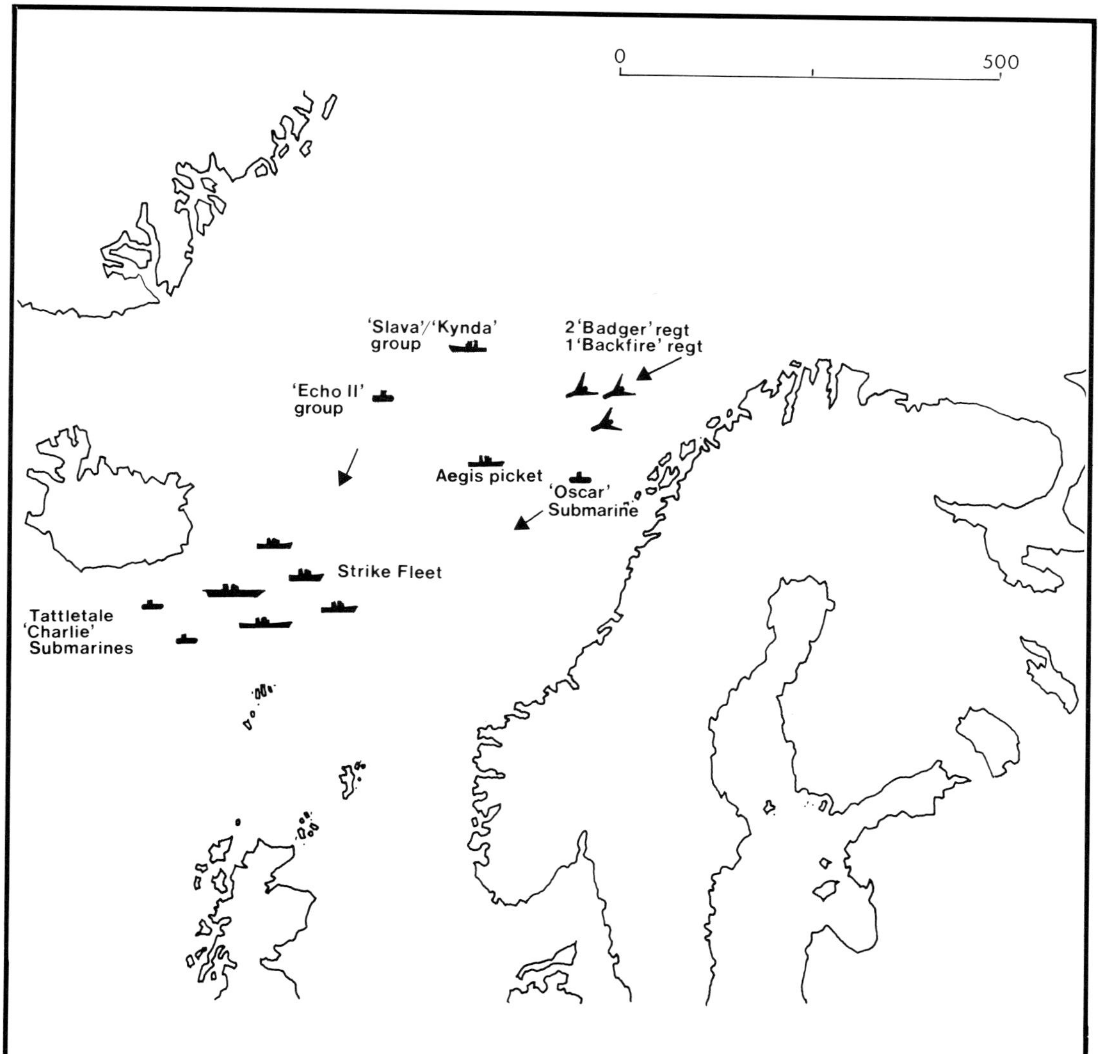

THE FIRST SALVO: the Plan

Engagement, but he is still empowered only to attack targets that show definite hostile intent. However, when the moment comes, he is helped by circumstances.

Far to the north, a Soviet task group headed by a 'Kirov' class battlecruiser is operating, apparently in support of Soviet SSBNs. It is being shadowed intermittently by S-3s from the US carriers, an unaccustomed role for them but made logical by the plethora of shore-based aircraft available for their more usual ASW mission. In the Arctic twilight the S-3 on task approaches the outliers of the Soviet force and — on whose authority is not clear — is engaged and shot down by a SA-N-6 missile. The Mayday message, with a clear indication that enemy action is responsible, is intercepted and this is enough for the shore authorities to relax Rules of Engagement to allow attack on any aircraft not identified as friendly.

The action is also enough to push the Russians over the brink. While holding for a little longer their invasion of northern Norway, for which their land forces are in readiness, they receive Defence Council agreement to mount attacks on NATO forces at sea in the area to the north of the Gaps. So they have approval for the first salvo without ideal conditions in which to launch it.

The co-ordinated assault which is planned includes regimental attacks by two 'Badger' and one 'Backfire' regiments; a SS-N-19 attack by an 'Oscar' class submarine which has worked undetected into the grain of the CVBGs; two

Elements of the Soviet Northern Fleet

Left:
The aircraft carrier *Kiev*.

Below left:
The battlecruiser *Kirov*.

Right:
An 'Echo II' class submarine.

Below right:
A 'Charlie' class submarine.
All Crown Copyright

close-range missile attacks by 'Charlie' class submarines, each a tattletale on one of the American carriers; an attack by SS-N-3 and SSN-12 at extreme range from a hastily-assembled group including one 'Slava' and one 'Kynda' class cruiser, closing from the northeast; and one by similar missiles from a better-placed group of three 'Echo-II' class submarines south of Jan Mayen Island. It is less, much less, than the Commander-in-Chief of the Northern Fleet had hoped for; but he is grateful that at least his plea for redeployment of the 'Backfire' regiment from the Baltic to the Northern Fleet was granted.

COMSTRIKFLT has resource problems too. He can thicken up his CAPs to an extent, but dare not over-extend his AEW, so still has only two such aircraft airborne. There is some flank guard from elderly British AEW Shackletons (British AWACS is not yet available) but their facilities for controlling fighters are minimal by US standards. As for radar emissions, here is a real quandary. He opts for AEGIS transmission from the two flank ships as well as the picket, with the main force continuing to rely on passive means until attacks develop.

When the attack does get under way, inevitably some things go wrong for the Soviets. Three target indication 'Badgers', having been sent out in advance of the regimental attack, fall foul of the reinforced CAP and are shot down. The fourth is incapable of providing all the necessary information; a 'Bear', detailed to work with the surface force and the 'Oscar' class submarine, is diverted to help. In this unexpected role it does its best but neither the aircraft, the submarine nor the surface ships get the easy and assured information to which they were accustomed in exercises. A second 'Bear', working to the westward in support of the 'Echo-IIs', is unaffected and can concentrate on its task of providing target information.

The air dispositions made by COMSTRIKFLT turn out to have been broadly correct. The 'Badger' strikes do develop along the threat axis, and both regiments are intercepted before reaching their break-up point, where they planned to diverge in order to attack from different bearings. In consequence a high proportion are shot down and only a remnant survive to fire their missiles. The 'Backfires' are another matter. Coming in behind the 'Badgers' and with their superior speed, they are largely unscathed by the CAPs which are now low on ammunition, and the great majority of their missiles are fired, from bearings up to 70° apart. Co-ordination is by no means perfect but most missiles are fired to reach the target area within the space of a few minutes.

Co-ordination is also achieved by the 'Oscar' class submarine. Helped not only by the 'Bear' but by its modern information processing systems, it gets its missiles away to arrive in the same time bracket as those of the 'Backfires'. The same does not apply to the surface force. Butting at high speed into the weather, its missile firing is both late and spasmodic; only half the missiles get away, and those at lengthy intervals. The 'Echo-IIs' also have problems with weather; they do however get a dozen missiles away, and their co-operating 'Bear', in a set-piece situation, has provided accurate information.

The warning given to the American force by the early appearance of the reconnaissance 'Badgers' helps them to dispose of one of the 'Charlie' class submarines before it can fire. There had been intermittent contact on this submarine for some hours before, enough to provide a datum position on whose basis it could be relocated and attacked. The other 'Charlie' class remains undetected before firing, but does not achieve co-ordinated timing with the main attack because of its need, in the face of heavy surveillance, to manoeuvre into a firing position.

The missiles fired therefore consist of some 25 air-to-surface missiles, mostly AS-6 'Kingfish', and 15 SS-N-19 submarine-to-surface missiles, all timed to arrive at the two highest-value targets (the US carriers) in the space of five minutes; and a ripple of subsequent attacks from weapons as diverse as AS-5 'Kelt', SS-N-3 'Shaddock' and the short-range SS-N-7 'Siren', spread over a couple of hours.

If the Soviets had had fully accurate information on the disposition of the Strike Fleet this would have been an extremely serious weight of attack. However, the cautious emission policy and dispersed formation decided on by COMSTRIKFLT have to some extent paid off. The Soviets had correctly identified the western of the carriers but misread the composition of the eastern group. In consequence, the greatest weight of targeting in the concentrated five-minute assault falls on the outermost AEGIS picket, the western of the two carriers, and a destroyer (fitted with an echo enhancer) in the eastern CVBG. In each case there is rapid engagement of the threat by Standard missiles, particularly from the AEGIS ships. Nevertheless, several missiles penetrate to the close-in weapon envelopes of their targets. Here they are subjected to Vulcan-Phalanx engagement from the carrier, Sea Sparrow from the destroyer and gunfire from the AEGIS cruiser; and all three use jamming and chaff. The AEGIS picket and the destroyer are nevertheless overwhelmed and sunk, by two and one missiles respectively. The carrier sustains one hit which restricts its deck operations but still allows it to steam.

The US force has largely ridden out the first salvo, but the follow-up ripple presents its problems too, not least in the high level of false alarms in the force as crews become accustomed to the realities of missile threat. Expenditure of ammunition is in some cases prodigal. The SS-N-7 attack from the surviving 'Charlie-II' has some success in badly damaging a frigate; and the underway replenishment group unfortunately gets in the way of the Soviet surface force's missiles, resulting in the abandonment of an oiler. But the most serious outcome of this phase is further damage to the western carrier from the 'Echo-IIs' rather old-fashioned SS-N-3 attack which, coming from well to the west of the threat axis, largely outflanks the outer layers of missile defence. Two more hits seriously restrict this carrier's ability to fly for 24 hours.

This has several effects. COMSTRIKFLT revises his timetable for proceeding to the operating area in the north, and calls on his third CVBG to use utmost despatch to join him north of the Gaps. This entails its ceasing flying operations almost completely. However, at this point COMSTRIKFLT has cause to be grateful to the RAF. Well placed in the Scottish airfields, it can give fighter cover to the third CVBG as it goes through the Gaps, and indeed to the stricken western carrier during its enforced downtime. Procedures to avoid blue-on-blue, well practised in the past, pay off now.

The British ASW carrier group, still operating to the eastward, has so far remained unscathed. But its luck is beginning to run out, for the Baltic Fleet Air Force is coming into the act. Sea Harrier CAPs, now being flown, and helicopter AEW are no match for 'Backfire' attacks launched from the Kaliningrad area and overflying Norway. Sea Dart and, more particularly, Sea Wolf have good success against the ensuing missiles, and guns, jamming and chaff see off several more; but damage to the force is severe. The basic problem here has been that COMSTRIKFLT could not cover all his flanks with AEW and CAP, and in spite of its efforts to get under the US cover the British force was still exposed.

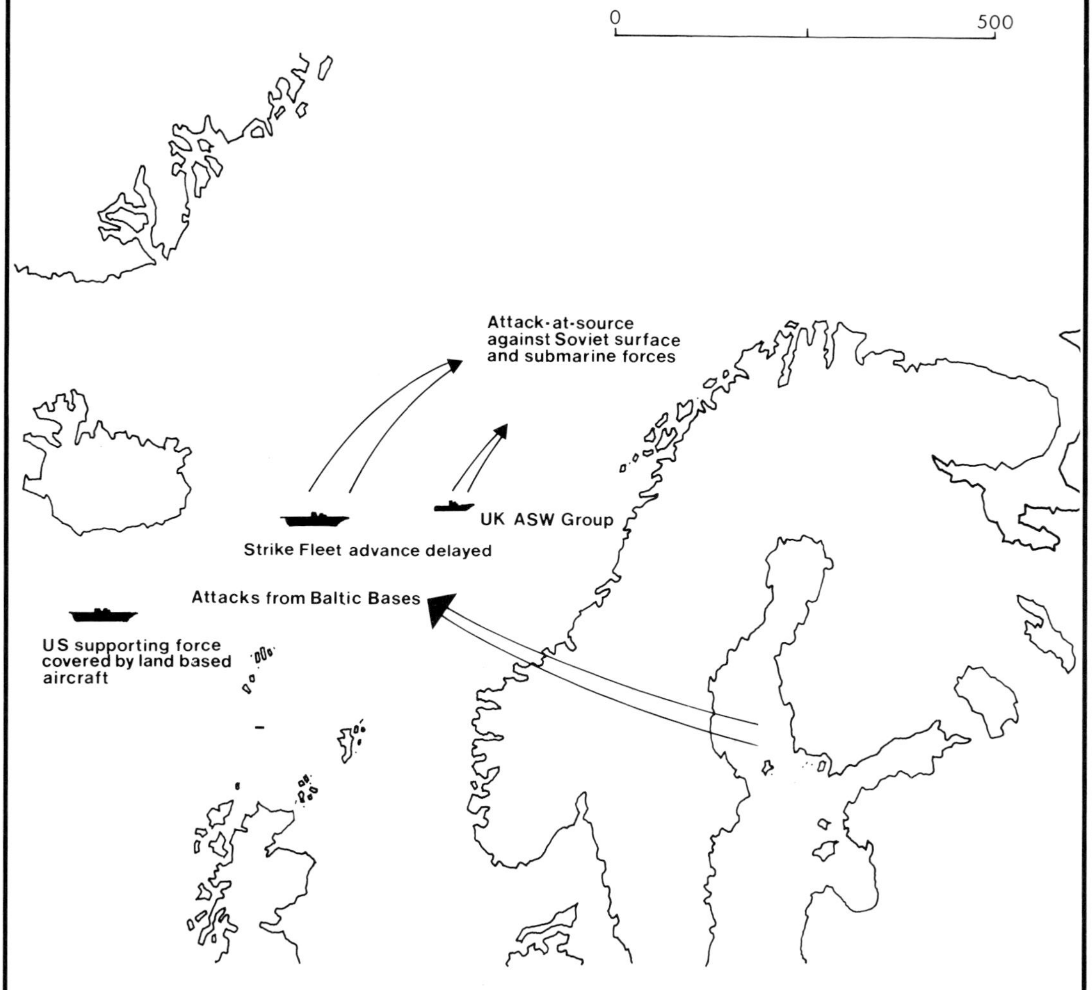

SUBSEQUENT DEVELOPMENTS

As it turns out, this is the low point of the campaign. COMSTRIKFLT's reinforcement is successfully accomplished, he is able to turn to attack-at-source not only against Soviet forces at sea but against the Kola base, and British forces receive reinforcement in their turn. The air defence situation remains tense, not least in terms of ammunition state, but control is regained.

There we leave the story, but one or two final points must be made. First, the action has been deliberately simplified and the forces engaged on both sides assessed as a good deal smaller than orders of battle (and, no doubt, the plans of the two Alliances) indicate. Mainly, this was done to illustrate in a reasonably understandable form how the principles of air defence are applied in a complex, all-arms situation. But partly, also, it was done to suggest that set-piece actions with maximum concentration of force are the exception rather than the rule, and that both sides may have to accept engagement in situations that are far from ideal for them.

Secondly, some forces in the order of battle on either side have not been inserted. The influence on the Soviet side of coastal missile-boats of the 'Osa' or 'Nanuchka' classes, the chances of a move out of the Baltic by major Baltic Fleet surface units in a period of tension, and any possible move by the 'Kiev' class aircraft carriers, were left out of the account. On the NATO side, there has been no

CAC
300

Left:
The S-3A Viking reconnaissance aircraft, an important element when the Strike Fleet turns to attack-at-source. *US Navy Official*

Below left:
An A-7E Corsair strike aircraft. *US Navy Official by Charles E. Solseth*

Right and below:
Logistics will be a prime consideration in any Norwegian Sea scenario. Shown are reloading a Sea Wolf and refuelling at sea. *Crown Copyright/ US Navy Official*

mention of the Standing Naval Force Atlantic, nor of the way in which the recommissioned battleships of the United States Navy might be used either to escort amphibious forces or to draw down fire which otherwise might descend on the carriers. Finally, it has not been possible to discuss the part that can be played by NATO E-3 AWACS aircraft: a topic that bears not only on the Norwegian Sea scenario but on the immensely complex business of producing an integrated air picture throughout the Alliance area. All these were left out for simplicity, but they need mention here.

Finally, it was assumed that only conventional explosives were used on both sides. This is the most questionable assumption of all, for it has been implicit in Soviet policy and planning until quite recently that early use of nuclear munitions was to be expected. Had this occurred in our scenario, ship casualties would have been considerably greater initially but, on the other hand, the virtual immobilisation of the Kola base would have quickly followed.

A Medium-Power Encounter

The outline story of the British campaign in 1982 to recover the Falkland Islands from Argentine occupation is too well known to need repetition here. This account will deal only with the air defence of ships in that campaign. However, it is worth making one or two general observations before the narrative starts.

The Argentine Air Force and Naval Air Arm aircraft available to attack British ships outnumbered the available British air defence aircraft by about six to one. But most of the Argentine aircraft were shore-based, so the advantage of mobility lay with the entirely ship-based British; subject to operational necessities, they could operate their ships at the extremes of Argentine unrefuelled aircraft range.

The trouble was that the operational necessities included (once all diplomatic effort had failed) an opposed landing on the Falkland Islands by substantial ground forces; and the Falklands were indeed within the range of Argentine aircraft operating from shore bases. Such operations had always been thought to require full control of the air above the landing area and its offshore approaches.

The problem before Adm Woodward was therefore a delicate one. In the pre-landing phase, which was likely to last several weeks, he had to draw out the Argentine Navy and Air Force and, manoeuvring to his best advantage, establish supremacy or at least dominance over them; but at the same time he had to preserve the major elements of his force, since it was the only source of combat aircraft for the post-landing phase.

The First Day

Hostilities opened on 1 May. It is clear from accounts written in hindsight that the impression given by the media to those at home — that the Argentine effort on that day was confined to 'probing' operations subsequent to the RAF's Vulcan bomber attack on Port Stanley airfield in the early hours — was inaccurate. The Argentine Air Force launched over 50 sorties that day, the first series to contest the airspace around East Falkland and the area to the east of it where the British fleet was operating, and the second directly to attack ships.

The British air defence effort was in action early. In spite of the need to mount a follow-up attack on Stanley airfield by Harriers soon after dawn, a CAP from 801 Squadron was flown apparently from about the same time, and fighter effort was maintained all day. The first Argentine sortie was from Mirage aircraft coming in high, attempting to draw the Sea Harrier CAPs up to the Argentines' best fighting altitude. This gambit was

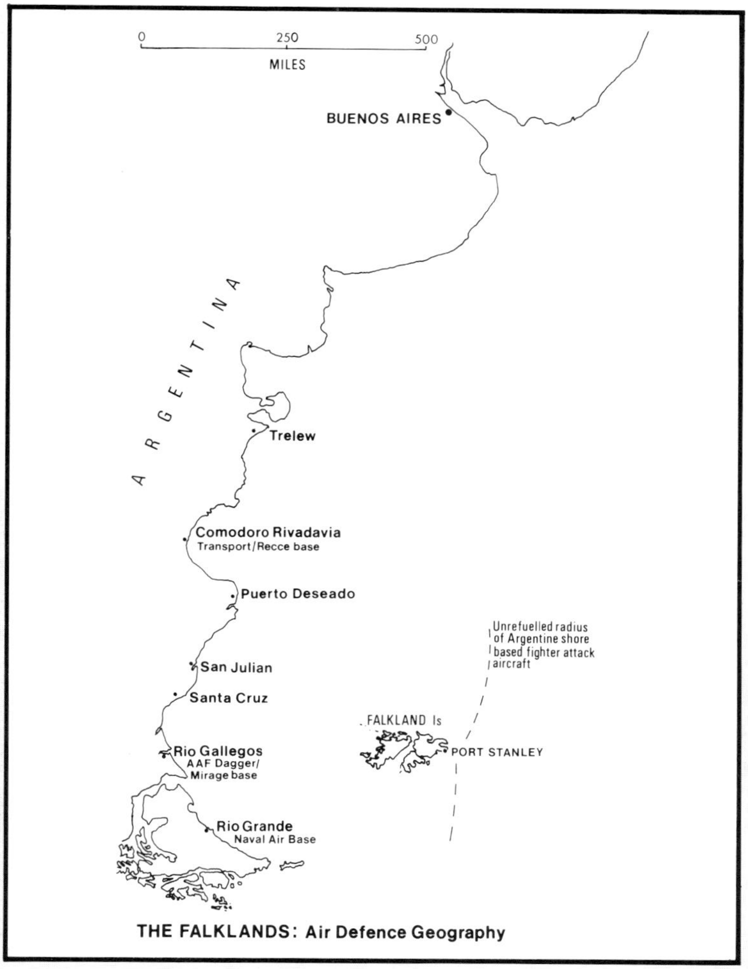

THE FALKLANDS: Air Defence Geography

Far left:

The influence of armoured ships, like the newly-recommissioned battleship *Iowa*, may be considerable in a Norwegian Sea context.
US Navy Official by PH1 Jeff Hilton

Below:

The sea/air equivalent of hand-to-hand fighting: an Argentine Mirage aircraft overflies a British frigate in Falkland Sound.
Press Association

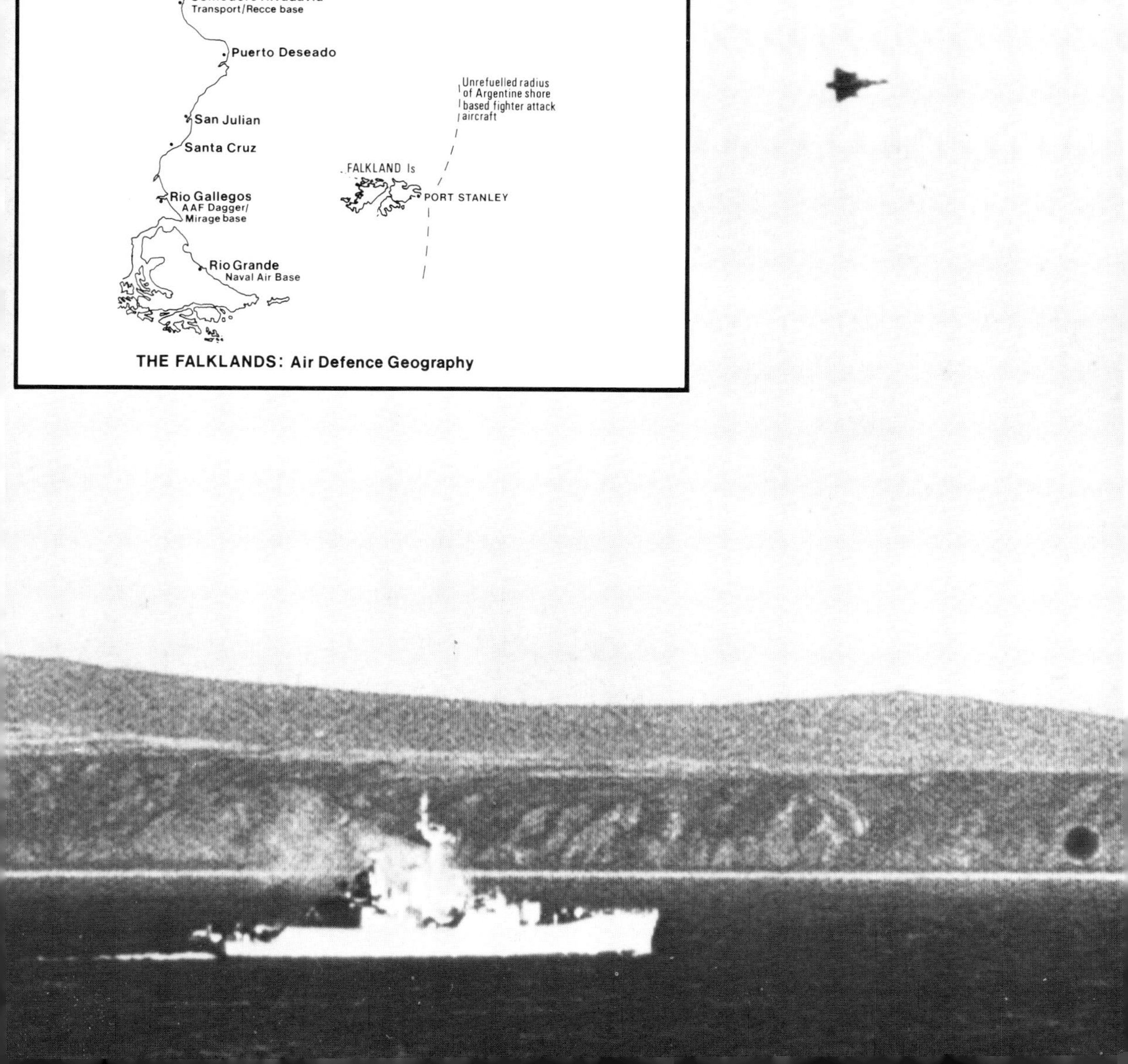

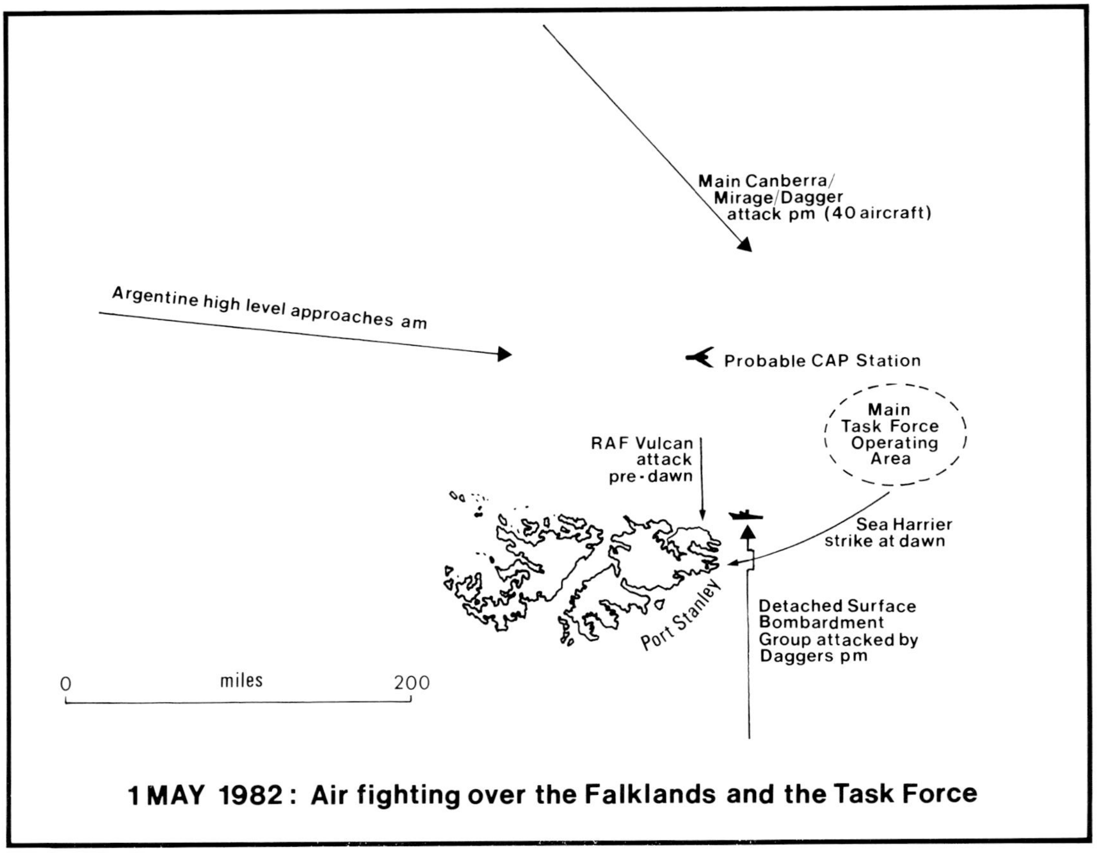

1 MAY 1982: Air fighting over the Falklands and the Task Force

Left:
Sea Harrier and ski-jump: two essential elements to Falklands success.
Crown Copyright

Right:
A 'Conqueror' class nuclear-powered submarine. Attack-at-source on the *General Belgrano* neutralised the air threat from Argentine surface-to-surface missiles. *Crown Copyright*

declined, no doubt largely due to the good judgement of the pilots and controllers but also perhaps because the British height-finding capability was restricted; there is a tale of a 'merged plot' on the flagship's radar where, indeed, British and Argentine aircraft were occupying the same bit of sky horizontally, but some 20,000ft apart in the vertical plane.

Later in the day there were several engagements between aircraft of the two sides. The first was an ineffective attack by an Argentine aircraft with its French air-to-air missile on a Sea Harrier which reported it as an Exocet; this caused a flurry in the main force. But the main effort occurred in the afternoon when a mixed Argentine force of about 40 aircraft approached the fleet, the Canberra bombers attempting to fly under the lobe of the ships' air warning radars. They were unsuccessful in this and interception was achieved, one Canberra being shot down and the rest breaking off the attack. Meanwhile, other attacks by faster aircraft had developed against a group of ships bombarding the shore. *Glamorgan*, leading this group, took control of the CAP and directed them on to a group of Argentine Mirage aircraft, one of which was shot down and a second damaged. The latter was shot down by Argentine anti-aircraft fire when attempting to make a forced landing at Stanley — a classic and tragic example of blue-on-blue.

A follow-up attack by a group of Dagger aircraft on the bombardment group was, however, initially unopposed by CAP and thus was subjected only to the inner layers of air defence. In this particular group they were pretty thin: the ships were *Glamorgan* (Sea Slug and Sea Cat), *Alacrity* and *Arrow* (Sea Cat only), though all had 4.5in (114mm) guns. The aircraft fired cannon and dropped bombs; all the bombs missed but the cannon fire caused minor damage. So far as is known, the aircraft were undamaged by the counterfire, and the only air defence missile fired (a Sea Cat) missed. The bombarding group moved away from the shore towards the main force. A second group of Daggers was attacked by the Sea Harriers and one aircraft was shot down.

Thus the first day of air defence ended satisfactorily for the British force — indeed subsequent information suggests that it was even more successful than was thought at the time, since the Argentine Air Force apparently decided after that experience that a battle of attrition in the air was an incorrect policy. In any event, the second day was quite different and brought out other principles of air defence at sea.

2 May: Belgrano

Adm Woodward has given a public account (long before the political rumpus over the *Belgrano* gathered momentum) of his concern on the evening of 1 May that the Argentine carrier *Vienticinco de Mayo* had slipped past the British submarine detailed to trail her, while the *Belgrano* might, by going over the Burdwood Bank, do the same to HMS *Conqueror*. The two Argentine forces, one to the northwest and the other to the southwest of the Exclusion Zone, constituted an air threat to the British force with several elements: A-4 aircraft from the carrier, and shipborne Exocet missiles from the carrier's escorting destroyers and from the *Belgrano*. To fire Exocet the Argentine ships would have to penetrate the Exclusion Zone, but they would not need to do so to mount an air strike.

It is now known that the Argentine carrier did indeed prepare to launch an A-4 strike at dawn on 2 May; the aircraft were ranged and bombed-up, and only lack of wind over the deck prevented their launch. In the meantime, Adm Woodward had requested, and received, permission to relax Rules of Engagement in order to attack surface ships outside the Exclusion Zone. He was unable to do this against the carrier force, whose position was uncertain and which had no submarine in contact. But, as is well known, the *Belgrano* was torpedoed and sunk by *Conqueror* at about 7pm that day.

It was an application of the principle of attack at source, and, as it turned out, its only major application during the conflict. For, faced with this demonstration of the potency of the British

nuclear-powered submarines, the Argentine fleet retired to local waters and did not try conclusions again.

So far as direct air defence was concerned, the British force had two quiet days on 2 and 3 May; the weather had more to do with it than the shock of the *Belgrano*, in all probability. For on 4 May came a serious challenge to the British air defence arrangements, and their first real failure.

4 May: Sheffield

The Task Force was, according to all accounts, about 70 miles southeast of East Falkland in a standard formation, with no ships detached for inshore bombardment. The carriers (*Hermes* and *Invincible*) were the easternmost ships, each being attended by a Type 22 Sea Wolf-fitted frigate to give close-range missile protection. A few miles to the west were three Royal Fleet Auxiliaries; some say these were put there as a physical barrier against missiles. If so it was an understandable disposition in view of the overriding importance of the carriers. Further west again, some five or six miles, were three frigates and a destroyer on a screen which had both air defence and anti-submarine purposes. And, finally, the three available Type 42 Sea Dart ships were in picket stations some 20 miles from the carriers, spread through a sector of about 90° with its axis (which may well have been designated a threat axis) to the westward.

Ships' companies were at defence stations, a state in which half the crew are closed up for action and the other half off duty. This is a state which can be maintained almost indefinitely in conflict, since it allows men enough sleep and relaxation, while those closed up are capable of initiating and conducting all the actions necessary to react to quickly-developing threats. The training of all British ships is done on this basis, although naturally, once a threat is known to be acute, the ship's company is sent to action stations where the top level of expertise and multiple target handling is to be expected.

The electronic emission policy of the Task Force almost certainly included liberal use of radar by the pickets, since the Argentine information on the force's whereabouts came from an ESM aircraft which detected a Type 42's radar characteristics. A force of two Super Etendard aircraft, each carrying one Exocet AM39 missile, was launched from a shore base and, refuelling in flight, proceeded to attack the British force.

At about 2pm HMS *Sheffield*, the southernmost of the Type 42 pickets, was transmitting routine messages to a communications satellite, which

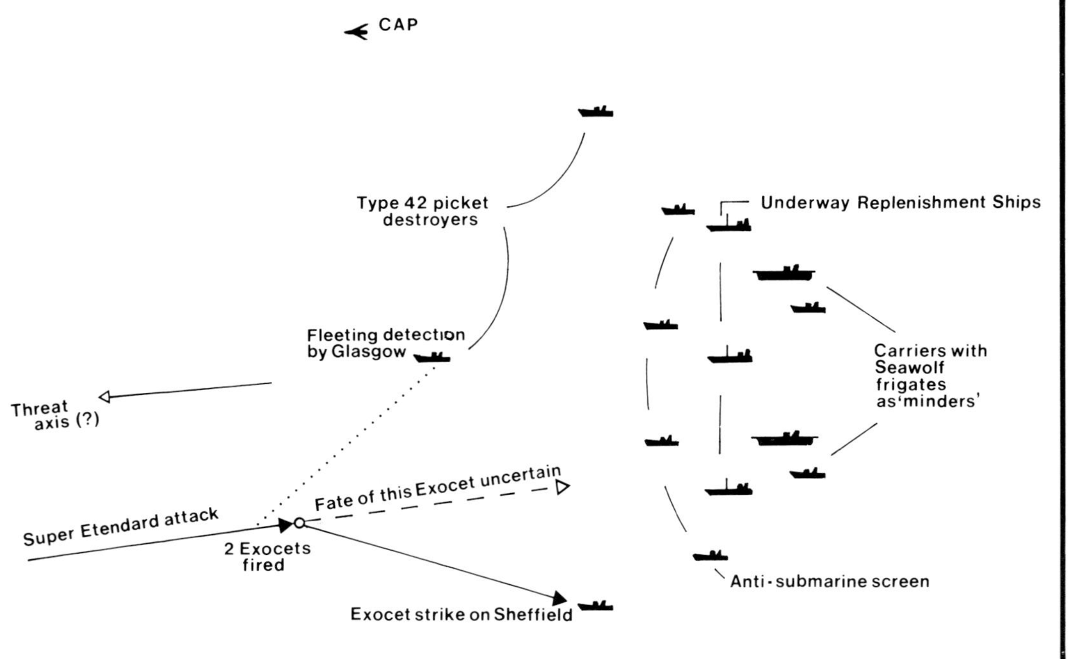

SUPER ETENDARD/ EXOCET ATTACK ON THE TASK FORCE, 4 MAY 1982
Exocet was probably released about 30 miles from carriers

meant that her ESM equipment was operating at greatly reduced efficiency; it is possible that her air warning radar was also affected. To the north of her, HMS *Glasgow* made a brief detection of a Super Etendard radar and, in accordance with the standard procedures, fired chaff for self-protection and at the same time alerted the rest of the force.

But time was not on the *Sheffield's* side. She was much closer to the firing aircraft than *Glasgow*; it is quite probable that the missiles were already on their way when the detection was reported. At any event, as is well known, she was unable to engage the missile approaching her by any countermeasure, was hit and set on fire and subsequently sank. The other missile made no hit; it may well have got beyond the inner screen before running out of fuel and falling into the sea.

A number of air defence lessons come out of this affair. First, there was a critical deficiency in the equipment of the British force: it had no AEW. Because of the criterion, imposed by governments since the late 1960s, that equipment must be justified by its necessity for operations in support of NATO, the Royal Navy had not been allowed to re-provide the organic AEW lost with the fixed-wing aircraft carriers. After all, it was argued, in the Eastern Atlantic all the AEW needed was provided by shore-based aircraft or by the US Strike Fleet. But this was no good in the South Atlantic, and the Task Force paid dearly. Had even helicopter AEW been available, it is almost certain that the Etendard force would have been detected in time to vector the CAP — which was airborne and well placed — into an attacking position well before the missiles were fired.

Secondly, the lesson was relearned that the amount of time available, once a missile has been fired, is very short indeed. Reaction has to be measured in split seconds. This is, of course, the reason for constant training of ships' companies, in both the action and defence states, in the so-called 'Zippo' procedures — those drilled reactions that are a response to detections of approaching menace.

Finally, and this was a credit to the British side, the formation of the Task Force made it difficult for a missile to penetrate along the threat axis to the prime target before it — and, indeed, its parent aircraft — could be engaged. Had the Etendards' attack been pressed home to the point where the missiles had a good chance of hitting the carriers, they would almost certainly have come under heavy fire. It may be, of course, that they had only a hazy idea of the formation of the Task Force; this would be consistent with the fact that they used ESM information. If so, there had been a partially successful denial of information to the Argentines: another application of the principles of air defence.

Above:
Sea Dart, whose capability (well known to the Argentines) had the effect of driving attacking air forces down to the deck. *Vickers Shipbuilding & Engineering*

Below:
Basic point defence: a sandbagged machine gun position on HMS *Fearless*. *Crown Copyright*

21 May: Bomb Alley

The events of 21 May, the day when British troops landed in force at San Carlos, may well go down as unique in history as the only occasion on which such a landing force's air support was heavily outnumbered by opposing air forces. For, in the days between the loss of the *Sheffield* and the landing, only sporadic air action had taken place, the Argentine Air Force was still 'in being' and the Naval Air Arm was prepared to operate from shore bases.

The day was unusual, as well, in that most of the smaller British warships were 'on the gunline' in confined waters with the purpose of protecting the amphibious units, and in these almost landlocked areas their air warning radars were of little help. Once again, AEW would have transformed the situation. In its absence, control of the CAPs (two, later three, two-aircraft units were kept airborne) was given to *Antrim* and later *Brilliant*, whose air warning radar and aircraft directors seem to have performed particularly well.

The carriers, even on this day of all days, had to be kept to the eastward as far as possible out of harm's (and Exocets') way. This meant, of course, that the Sea Harriers' time on CAP duty was cut down by their need to transit to and from the carriers, as well as their reliance on forward-placed ships for air direction. Nevertheless, during the day this outer layer of air defence accounted for some eight Argentine aircraft, a high loss-rate observing that only 50 or so sorties from the mainland were mounted. No Sea Harriers were lost in return; throughout the conflict their skilled handling, agility and Sidewinder missiles gave them the advantage.

The inner layers of air defence, deployed under unfavourable conditions, were severely tested. Ship-to-air missiles optimised to open-sea work were not of great use in these almost landlocked waters, though the fear of Sea Dart's capabilities against medium-height aircraft did at least drive the Argentines down to the deck where their bombs had fuzing difficulties. But guns, and particularly close-range weapons, were the flavour of the day, and ships did their utmost to maximise the number of channels of fire, with improvisations from General Purpose Machine Guns on upwards. They may not have shot down many aircraft, but they must have acted as a deterrent to accuracy and helped ships' morale no end.

The ordeal of HMS *Ardent* stands to typify this torrid day under the most difficult of air defence conditions. The southernmost ship of the 'gunline', which was the best formation that could be devised for these restricted waters, she was early in action to bombard Goose Green. After that she was subjected to day-long attack. Her 4.5in gun and Sea Cat short-range point defence missiles were supplemented by small arms, including a GPMG operated by the NAAFI Manager. According to one account she took 17 hits from bombs and missiles. (Exocet, of course, was not among them; the land clutter did at least preclude the Argentines' use of this lethal weapon.) This estimate may be an exaggeration, but there is no doubt that she sustained very heavy punishment. Ablaze aft, her Sea Cat destroyed and listing severely, she was towed for a time but eventually her back broke and she sank in Falkland Sound. Twenty of her company were killed.

It would probably be wrong to say that *Ardent* was singled out by the Argentines in the sense that they were out to get one ship for certain that day. Several other destroyers and frigates were struck by bombs, many of which did not explode; and *Antelope*, not as lucky as some, was lost when one of these later did detonate. Nevertheless, during the day, the *Ardent* did seem to have become a symbol for both the British and Argentines and her determination and endurance were in the Royal Navy's highest traditions.

It may have been this sturdiness that caused the Argentines to pause on 22 May. I can find no explanation in the literature for this lack of activity; it does not appear that the weather was particularly unfavourable. Certainly, to the listener at home it looked as though, if the British force could get through one more day, all would be well. So it proved, and it was as well that the Argentines did not continue their attacks on that day, for the British force was hard pressed. Indeed, throughout the conflict there seemed to be a lack of purpose in the Argentine High Command that resulted in piecemeal attacks and success not followed up. It was just as well. Nevertheless, there was one further day of success for them: appropriately their national day, 25 May.

Above:
Though transports were attacked, the main attacks in Bomb Alley on 21 May 1982 were directed at warships on the gunline.
Crown Copyright

Below:
HMS *Ardent*, lost on 21 May after attacks of great ferocity and sturdy defence. *Crown Copyright*

Right:
Sea Wolf, a generally successful point defence missile system in the Falklands.
British Aerospace Dynamics Group

25 May: Coventry *and* Atlantic Conveyor

A formation which had gained some popularity during the operation was the '22/42 combo', in which a Sea Dart destroyer and Sea Wolf frigate were put together, providing mutual area and point defence. Stationed in the northern approaches to Falkland Sound, their positive operational purposes were threefold: they could warn forces in the landing area of impending attacks; they could engage opposing aircraft either entering or leaving the landing area; and they could draw fire from both the landing area and from the carriers operating to the eastward.

On 25 May HM Ships *Broadsword* and *Coventry* were operating in this way. Their position quickly became known to the Argentines, and in the early afternoon an attack was mounted by four Argentine Air Force Skyhawks; six had set off, but

Left:
Ships of the Royal Fleet Auxiliary which, with ships taken up from trade, were essential elements of the Task Force. *Crown Copyright*

Right:
For operations of great reach, combat aircraft organic to the fleet are essential.
Crown Copyright
by T. J. Tierney

two had to return to base on technical failure. The attacking aircraft were detected in good time and the CAP, which was favourably placed, was in visual contact when it was called off by *Coventry* which engaged with Sea Dart. This engagement was unsuccessful and the Skyhawks came on, the first pair making for *Broadsword*.

At this point the Sea Wolf system, which up to that time had performed well in the Falklands, showed an unexpected limitation. Perhaps, as *Broadsword's* Captain subsequently said, it was confused by the wingtip-to-wingtip aspect of the targets. At any event, the expected engagement in the automatic mode did not take place and a change to manual operation could not be made in time. The aircraft released four bombs unscathed, and it was fortunate for *Broadsword* that three missed and the fourth passed *through* the ship without exploding.

Coventry had less luck. The attack on her came after that on *Broadsword* — long enough after, indeed, for the latter (whose missile system was undamaged) to shift target to *Coventry's* attackers. But again the Sea Wolf system was huffed, this time by a self-protection manoeuvre by *Coventry* which had the effect of breaking the lock of the *Broadsword's* guidance radar. The attack was pressed home and three of the four bombs penetrated deep into the ship, which capsized and sank in little more than a minute. Nineteen of her company were lost; in the circumstances it is astonishing that the number was not greater.

The other Argentine success on this day was also due to careful planning. It was an outflanking operation by two Exocet-fitted Super Etendards which, after in-flight refuelling, approached the main body of the Task Force from the north. They fired their missiles on radar information, no doubt aiming for the large ships in the middle of the group; it is clear however that they made their approach on passive sensors, only switching on active radar at the last moment. In fact it was a very well-conducted attack.

The aircraft were fleetingly detected by *Ambuscade* on the northern wing of the force, and the same ship correctly detected the firing of the missiles. All ships so fitted fired chaff, and it is almost certain that one missile was rendered harmless by this. The other, however, proceeded and struck the *Atlantic Conveyor*, a ship taken up from trade to transport aircraft and stores. This valuable unit had subsequently to be abandoned and sank some days later; 12 lives were tragically lost when she was hit.

Again, the events of 25 May appear to point to many air defence lessons. I say 'appear' not only because one must beware of being too wise after the event, but also because all the facts available to the decision-makers at the time may not yet have come to light.

But it *appears* that some of the lessons are these. First, if the outer layer of air defence is in position to cope with a threat, it should be allowed to get on with it. It will need a highly sophisticated multiple threat assessment to leave such a threat to one of the inner layers, and great confidence too in the inner layer's efficiency. This lesson is particularly applicable to medium-power encounters where the risk of saturation attacks is a good deal lower than in superpower conflict.

Secondly, planners must beware of basing anything on claimed single-shot kill probabilities. That figure for Sea Wolf is in excess of 80%, but on two occasions on this day, Sea Wolf did not even fire at targets well within its scope. We have not been told how many Sea Wolf missiles were actually fired during the Falklands campaign, but the assessed Sea Wolf kills (varying between three and five, according to which publication you choose) suggest that even on missiles fired, the success rate was somewhat below the figure previously claimed by designers and manufacturers. This is not to denigrate Sea Wolf, the best point defence missile so far put into service. It is, however, to say that the historical tendency to over-estimate the effectiveness of new weaponry has in no way changed with the advent of new technology. Allowance must be made for this in both armament and logistic planning.

Thirdly, the danger of being outflanked must always be borne in mind. In the Falklands, particularly observing the lack of AEW, a threat axis was virtually imposed on Adm Woodward; he could do nothing other than what he did. The Exocet attack, working round that axis, nevertheless did not achieve complete outflanking, and this was to the Task Force's credit.

Finally, it is notable that the *Atlantic Conveyor* had no chaff launchers. Had she been so fitted, she might have survived. If such systems as now designed are so complicated that they cannot be fitted quickly, then a simpler system for rapid assembly and fitting on all merchant ships in time of war ought to be devised.

The tale of air defence in the seas off the Falklands is nearly done. It is not proposed to discuss the sad business of Bluff Cove, a case perhaps of over-confidence and lack of communication rather than basic deficiencies in air defence arrangements.

The campaign was, in all, an air defence success under exceptionally adverse conditions.

Below:
A submariner who achieved a remarkable feat of arms in air defence: Sir John Woodward, seen here with Lady Woodward on his return from the South Atlantic in 1982.
Crown Copyright

Supremacy was never achieved, but significant dominance was, particularly in the areas of shipborne fighter power and attack at source on sea-based threats; and this dominance, with the sturdy resistance by all other aspects of the air defence framework, was enough to carry the business through. It was a remarkable feat of arms.

Some Other Aspects of Battle

The two accounts above — one factual and one, maybe, factional — have left out one or two significant aspects of air defence at sea, and for completeness these must be covered, albeit briefly.

First, neither was concerned with inshore work against threats from missile-boats. This is a specialised field demanding extremely rapid reaction and as good surveillance and intelligence as can be devised. It should, however, be remembered that the attacker has problems too. Take the case of a frigate put 'in harm's way' to make — say — a point about freedom of passage in waters claimed by some other state. An attack is mounted by an 'Osa' class missile boat firing four 'Styx' missiles: one misfires, one misbehaves in flight and is never seen again, two lock on. But the frigate, fitted with good ESM and Sea Wolf, manages to shoot down one missile and, firing chaff, manages to seduce the other one. It survives.

Now put the case of a two-'Osa' attack and reassess. Much will depend on the degree of co-ordination by the attacker. It seems to this author that there are two ways of organising air attack that are particularly dangerous to the defender: one is the saturation attack, with very large numbers designed simply to swamp the defences over a short span of time; the other is the follow-up attack, with a 'second wave' designed to take advantage of a defending force thrown off balance or half-exhausted by fighting off a previous assault. Many goals are scored that way. But both these types of attack demand good planning, command and control.

Above:
NATO units exercising off Portland. Concurrently with the line manoeuvres, an air defence exercise is conducted in the operations rooms. *Crown Copyright*

Right:
The air defence of merchant ships is a subject which may repay further study.
Crown Copyright

Not too far removed from the previous case is the battle conducted by light missile-armed forces on both sides — a fight, perhaps, between neighbouring states. Here the best example so far is the Arab-Israeli Yom Kippur War of 1973. The Israeli Navy, theoretically out-ranged, outgunned and outnumbered, nevertheless won by a remarkable margin variously assessed as 19-0 at its most optimistic to 17-2 at the other end of the scale. It has not to my knowledge been fully written up in published work, but it appears that the Israeli success was due to a combination of excellent intelligence (including signal intelligence), sensible command and control, maintaining the initiative including using attack at source, high operational use of ESM and ECM, and the use of every conceivable channel of defensive fire when under attack. Conversely, the opposition showed weaknesses in all these fields. One extension may be made to this list of virtues in suitable circumstances: the use of geographical and hydrographical features by forces with local knowledge to disadvantage and surprise the opposition.

Finally, some mention must be made of the air defence aspects of convoy. A missile threat, particularly if it involves release of missiles at long range, is dangerous to large groups of lightly-defended ships. An escort fitted with area defence systems, and a CAP even of limited numbers and control arrangements, may be of value particularly against medium-power threats. But in any kind of conflict — medium or superpower — the resources available for direct defence of supply convoys are going to be scarce.

It may seem paradoxical, but in my view these considerations militate against very large convoys of several hundred ships. The area taken up by such a concourse of ships, particularly at anti-nuclear spacing of a mile or more apart, is such that even the best of area defence systems would be unable to cover it; missiles fired 'into the brown' would be certain to find targets. And a CAP flown from platforms within the boundaries of such a convoy would have a long way to go before getting into the optimum station to defend it. A better solution from the air defence, as well as the anti-submarine, point of view appears to be the system of medium-sized (say 30 to 40 ship) convoys covered by support groups with both ASW and air defence assets, and routed through the same broad area, though not — heaven forbid — along the same exact 'sea lane'. A more complete description of this system appears in *Anti-Submarine Warfare* in this series.

For self-defence, merchant ships should be fitted with chaff launchers at least, and machinery should exist not only for rapid fitting but for instruction in its use. There must also be a market for a strap-on point defence system with automatic warning; there seems no reason why the fire-and-forget systems such as RAM, scheduled for introduction in the early 1990s, should not have a version for rapid fitting in emergency to a wide range of merchant ships. If, that is, the merchant fleets of the West have any ships left by then; *laissez faire* by governments, which ought to know better, has gravely weakened the shipping industries of the USA and of most European countries, Britain being one of those most notably affected.

Epilogue

At a fairly early stage of this book it was said that the Battle of the Philippine Sea, with its crushing victory for carrier-based air defence in the face of heavy air attack, had burnt itself into the consciousness of the United States Navy. The consciousness survives today in the confidence with which that Navy has taken up The Maritime Strategy: of carrying the battle, if deterrence fails, to the Soviet forces in the Norwegian Sea and Sea of Japan, and beyond to the Barents and Okhotsk Seas and the Soviet Fleet bases if need be.

There are two vital strategic questions that face not only US but NATO planners: whether the strategy is necessary and whether the confidence is justified.

On the necessity for the strategy, this book need not make a judgement, since that is not its subject. But it may be worth saying that some positive operations in the maritime sector will, in the event of conflict, be expected by the land and air forces in Europe, which will be hard-pressed; and that reinforcement of Norway is, as it has been since the start of the Alliance, a precondition for the defence of that country's territory. To dispute the control of the Norwegian Sea, therefore, is a strategic option that must be considered, and to achieve such control would be a very desirable strategic objective — if it can be done.

So, is the confidence justified? Again, this book can make only a partial and qualified judgement, concerned with the air threats and defences that have been its subject. In this field, the United States forces and particularly the United States Navy have emerged from the foregoing discussion as world leaders by a long way. They command different dimensions of air defence power at sea from any other forces, including those of the Soviet Union. Only the USA has comprehensive AEW, high performance fighter aircraft that are organic to sea forces, and the formidable air defence performance of the AEGIS system. There may be a softer underbelly in the shortage of systems to deal with sea-skimming missiles fired from short range; but attack-at-source abilities, including the ability to attack short-range missile-firing submarines, are diverse and powerful.

Nevertheless, the threat is severe and increasing, and Soviet geographical advantages great. Both saturation and follow-up attacks, identified above as the most dangerous in the air defence context, are much more possible for the Soviets in the Norwegian Sea (and the Sea of Japan) than in the open oceans. The USN certainly could not count on a no-casualties or even a low-casualties fight. Its commanders would be faced with many decisions on whether to concentrate or disperse, reinforce or withdraw. They would indeed be men who could 'lose the war in an afternoon'. The best assessment that can be made is this: if anyone can do it, the USN can.

Soviet surface forces would quickly find themselves desperately short of the outer layers of air defence. Though faced with a lighter air-to-surface threat than they themselves pose to NATO in that area, they would have to cope with surface-to-surface missiles in many modes from many diverse launch platforms. In spite of their numerous channels of area and point defence fire, they would suffer heavily.

There could be strategic implications here reaching even beyond the question — important enough in all conscience — of who dominates the Norwegian Sea. For, if the Russians were to perceive themselves losing a conventional contest, might they not take the nuclear route in this theatre? It is a question that has been applied to both East and West often enough in land scenarios; it is not irrelevant at sea.

What of the rest of the world's navies? It may be unfair to class them as minnows swimming round the two Tritons, but the simile is inescapable. There is no doubt that in an East-West conflict, all NATO forces that could would do their best to hurry in under the US umbrella. Those that could not would hope the bulk of the threat was being diverted to the Strike Fleet. They would also make maximum use of integrated NATO information systems, so that they could at least get warning of incoming threats. Land-based air support would be looked on as a bonus by most, with the Germans best placed for attack at source with their naval Tornado aircraft, and the British comforted by the dedication of some parts of the Royal Air Force to the maritime role. But for many it would be a case of point defence, quick reaction and training. The principles of layered defence would still apply, but in practice the layers would be uncomfortably few and thin.

Such medium powers can expect to do better if they pick on opponents their own size, or can choose the sea areas in which they fight against stronger enemies, or can in other ways limit the scope of their opponents' air attacks. But even in such cases they will have to choose tactics and dispositions to minimise their limitations. These are most likely to be in the fields of early warning, of the outer layers of air defence, and of overall command and control, and they will tend to sharpen as the distance from the homeland increases. Effective reach at long distances is in many ways more difficult to achieve in the above-water battle than in the underwater one. At such distances from home, warning and combat aircraft embarked in the force will be necessary for any operation that may go to the higher levels of conflict.

In this sudden, hard-hitting world the offence must have built-in advantages. For medium powers as well as superpowers, the attractions of attack at source, particularly pre-emptive attack, will be great. But the political situation, reflected in Rules of Engagement, may not permit it. Some governments may be confident enough, or reckless enough, to plan on the assumption that they can get away with pre-emptive action. Others will be constrained to plan so because that is the only thing their resources will allow them to do. But forces designed for such action only, without the defensive ability to take a counter-attack, will have no staying power. This can only be acquired by translating into materiel, training and organisation the principles of this most difficult and demanding art, air defence at sea.

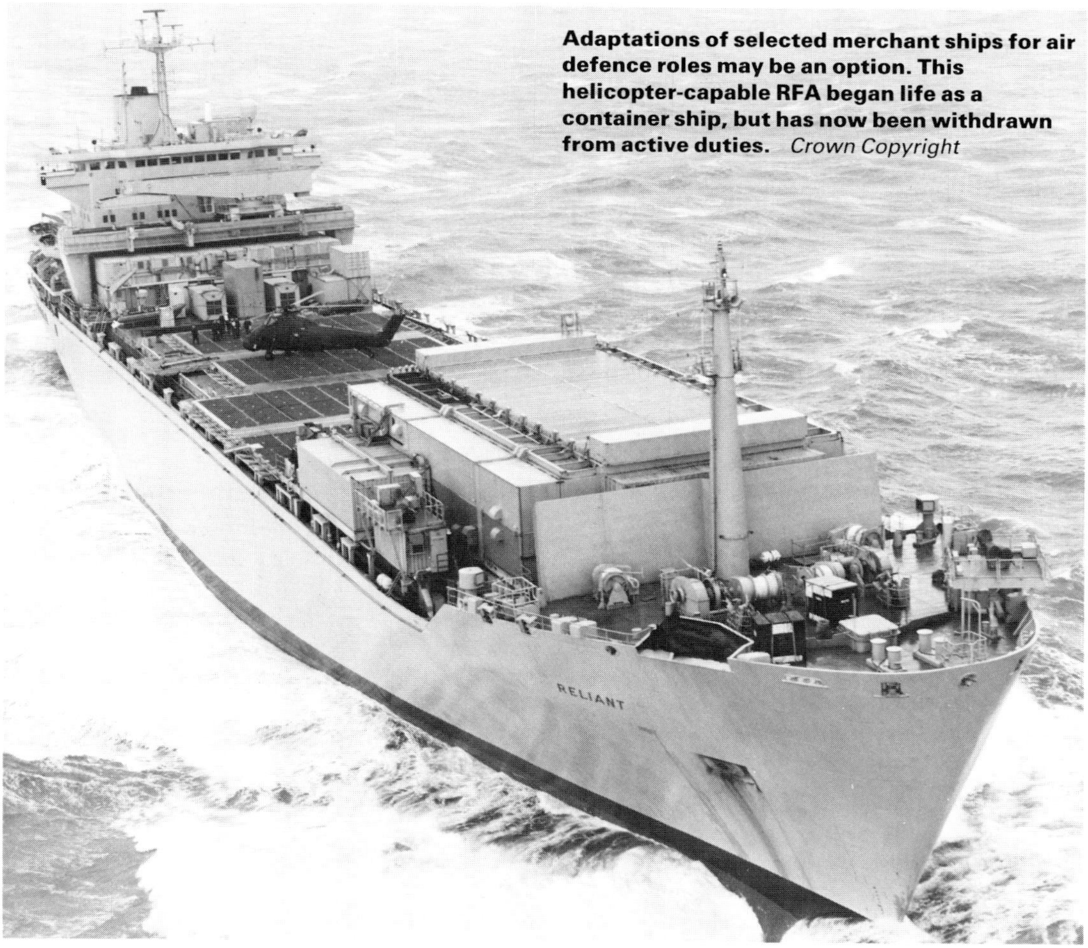

Adaptations of selected merchant ships for air defence roles may be an option. This helicopter-capable RFA began life as a container ship, but has now been withdrawn from active duties. *Crown Copyright*

Bibliography

On historical aspects, the main general and background sources were, as ever for the 20th century, Arthur Marder and S. W. Roskill: *From the Dreadnought to Scapa Flow* (OUP, 1961) and *The War at Sea* (HMSO, 1954) respectively are works of outstanding scholarship and authority. Vice-Adm Sir Arthur Hezlet's *Aircraft and Sea Power* (Stein & Day, 1970) is a valuable commentary. Early thinking was typified by an anonymous article in the 1913 *Naval Review* and by the work of F. W. Lanchester in *Aircraft in Warfare* (Constable, 1916).

On Soviet doctrine, Gorshkov's *The Sea Power of the State* (Pergamon, 1978) is still the most important single source, but later work indicating a more flexible approach to intermediate levels of conflict has also been noted, and Robert van Tol's reading of recent Soviet exercises have not been discounted, although I do not accept his interpretation completely. The work of James McConnell, Michael McGwire, Norman Polmar and Milan Vego must also be mentioned. Details of Soviet material were derived from many sources, all secondary and all published. Reference works in the Jane's series, *Combat Fleets of the World* (Arms & Armour Press), the International Institute for Strategic Studies' *The Military Balance* and Brassey's *Defence Yearbook* were all consulted. Where there were differences (there are not many) I tended towards the more conservative figures.

So far as non-Soviet threats are concerned, it is clear that one cannot generalise about doctrine and indeed many nations that are potential threats publish little on the subject. It is doubtful, for example, if Argentine forces had much doctrine on the subject of attacking a British Task Force. Non-Soviet threat material is well documented by the sources already mentioned.

On 'The Means', I had no access to classified information; but a good deal of experience in the field, and an eye for the relevant in published work, has I hope helped to achieve a reasonably accurate picture. Several written sources have been of great value. Adm James D. Watkins USN on 'The Maritime Strategy' in US Naval Institute *Proceedings*, January 1986, lets slip (deliberately no doubt) a carrier battle group disposition for air defence. The principles of layered defence are adumbrated in many pieces of work, and one must mention particularly the outstanding articles of Norman Friedman in many publications. US Naval Institute *Proceedings* articles by Col W. Hays Parks USMC and Lt Cdr Robert E. Stumpf USN were most illuminating about the Libya operations of 1986. Finally, at the suggestion of Anthony Preston I acquired the papers of an international symposium on *The Air Threat at Sea* sponsored by the Royal Institution of Naval Architects and held in June 1985. This contained much thought about tactics as well as diverse ideas about materiel, and was an excellent source for the discerning.

On materiel, the sources were of great volume and, often, physical weight. Manufacturers' brochures contain a mass of information, though this does not always extend to performance figures. Standard reference books, already mentioned, do contain performance figures, and sometimes these differ quite widely between one publication and another — more so, it may be said, for Western than for Soviet equipments; make of that what you may. I deliberately did not go into the question of Single Shot Kill Probabilities, which quite often are claimed by manufacturers; these are figures above all others which must be treated with great scepticism. Of other sources, magazines such as *International Defence Review, Naval Forces* and *Military Technology* contain immense amounts of technical information and, often, operational comment; the last two in particular were rigorously combed. Finally, I should say that one of the most difficult things to acquire was a clear view of how modern, sophisticated electronic equipments work. Even with a good knowledge of basic radar theory, I often found myself quickly lost in the opacity of technical language. I hope my interpretations are not inaccurate, and are pitched about right for the lay person's understanding.

Finally, the two battles were differently researched. The NATO encounter owes much to literature supplied by US sources, ranging from the Watkins/Lehman Strategy statements to Congressional Committee Hearings. The pattern of Soviet response was derived in part from Gorshkov, in part from US and NATO assessments over the years promulgated in official publications and magazines such as *NATO's Sixteen Nations*. On the Falklands the indispensable works were Ethell and Price's *Air War South Atlantic* (Sidgwick & Jackson, 1983), Martin Middlebrook's *Operation Corporate* (Viking, 1985) and Cmnd 8758 *The Falklands Campaign — The Lessons* (HMSO, 1982). Conversations with those who took part helped to clarify the issues and give the factors proper weight. Admiral Woodward's account of the *Belgrano* appears in the RUSI *Journal* of September 1983.